ROUTLEDGE LIBRARY EDITIONS:
SOCIAL AND CULTURAL GEOGRAPHY

Volume 13

HUMANISTIC GEOGRAPHY
AND LITERATURE

HUMANISTIC GEOGRAPHY AND LITERATURE

Essays on the Experience of Place

Edited by
DOUGLAS C. D. POCOCK

Routledge
Taylor & Francis Group

LONDON AND NEW YORK

First published in 1981

This edition first published in 2014
by Routledge
2 Park Square, Milton Park, Abingdon, Oxfordshire OX14 4RN

and by Routledge
711 Third Avenue, New York, NY 10017

Routledge is an imprint of the Taylor and Francis Group, an informa business

First issued in paperback 2015

British Library Cataloguing in Publication Data
A catalogue record for this book is available from the British Library

ISBN 978-0-415-83447-6 (Set)
eISBN 978-1-315- 84860-0 (Set)
ISBN 978-0-415-73362-5 (hbk) (Volume 14)
ISBN 978-1-138-97214-8 (pbk) (Volume 14)
ISBN 978-1-315-84825-9 (ebk) (Volume 14)

Publisher's Note
The publisher has gone to great lengths to ensure the quality of this reprint but points out that some imperfections in the original copies may be apparent.

Disclaimer
The publisher has made every effort to trace copyright holders and would welcome correspondence from those they have been unable to trace.

Humanistic Geography and Literature

Essays on the Experience of Place

Edited by Douglas C. D. Pocock

CROOM HELM LONDON

BARNES & NOBLE BOOKS
TOTOWA, NEW JERSEY

© 1981 Douglas C. D. Pocock
Croom Helm Ltd, 2–10 St John's Road, London SW11

British Library Cataloguing in Publication Data

Humanistic geography and literature. - (Croom Helm
    series in geography and environment).
    1. Local colour in literature
    2. English literature - History and criticism
    I. Pocock, Douglas, Charles David
    820'.9'32   PR149.L/

    ISBN 0-7099-0193-3

First published in the USA 1981 by
Barnes & Noble Books,
81 Adams Drive,
Totowa, New Jersey, 07512

ISBN 0-389-20158-8

Printed in Great Britain by
Biddles Ltd, Guildford, Surrey

CONTENTS

TABLES AND FIGURES

PREFACE

This volume is composed of thirteen original essays on the theme of geography and literature. It evolved from a special session organised by the editor at the meeting of the Institute of British Geographers in Manchester in January 1979. With the exception of Evangeline Paterson, co-author with her husband in a joint contribution, all the authors are academic geographers who value imaginative literature as a rich source for exploring the nature of the man–environment relationship, for which the 'hard' positivistic stance is inappropriate. Collected broadly under the theme of experience of place, the diverse contributions fall perhaps into three main groupings according to whether emphasis is on the nature of imaginative literature as a source, insight into particular experiences, or illumination of particular places. The deliberate perspective applied to imaginative literature in this collection will, it is hoped, be of value not only to human geographers but also to students of English literature.

As editor of the volume, I acknowledge my gratitude to those fellow geographers who approved of the project and supported it by kindly contributing essays. I am also heavily in debt to Edith Pocock for producing the bulk of the typescript. Last, but not least, acknowledgement is given to both written word and lived environments that have provided, and continue to give, both enlightenment and nurture.

Douglas Pocock
Department of Geography
University of Durham

1 INTRODUCTION: IMAGINATIVE LITERATURE AND THE GEOGRAPHER

Douglas C. D. Pocock

Imaginative literature, a relatively small subset of the vast heterogeneous field of the printed word, has recently been espoused — 'used' would be an inappropriate description — by a growing band of geographers seeking alternative perspectives and insights in the study of man—environment relationships. Disillusioned by an era of logical positivism, maybe shell-shocked by the quantitative revolution, perhaps rediscovering the literary heritage of geography — whatever the reason, the realm of literature has attracted increasing attention from our eclectic discipline. For such engagement to be profitable and lasting, there must be at the outset a proper appreciation of the two participants.

Geography and Literature

Literature is an artistic creation and not a scientific construction, while geography — or human geography — may be either an art or social science. The latter, however, is acknowledged as an inappropriate stance from which to engage literature. Objective techniques, such as content or structural analysis, for example, cannot capture the essence of the created word, even though the 'data' produced may suggest new aspects of interpretation.[1]

The primacy of literature and the holistic nature of literary revelation are also acknowledged by the geographer, which may prompt the question whether the latter is not engaged on some secondary pursuit analysing the imagination of others rather than acting more creatively himself. The answer to such musing must be no, since, if literature is universal and speaks to the human condition, it cannot belong exclusively to students of literature. Literature illuminates all, and many disciplines concerned with man make use of its insights.

Geography, of course, has traditionally been the discipline without exclusive source material, engaged in a 'borrowing' exercise in which a distinctive and integrative viewpoint is brought to bear. Acknowledging the complementary field of literary criticism alongside creative literature, the geographer's perspective here seeks primarily to enrich his own experience and understanding, and only secondarily to add perhaps to the total appreciation of the art form. There is no intention of suggesting that there is a distinctively geographical approach to literature which literary critics ignore at their peril. Any such attempt could be scuttled by the analogous argument which Lewis used to refute a distinctively anthropological approach.[2] At the same time, however, it must be emphasised that the geographer cannot avoid bringing, even subconsciously, his or her own particular academic experience into the realm of imaginative literature, and that in this process the aim is in fact akin to that of literary criticism, in that any analysis is for the ultimate benefit of greater synthesis. To reiterate, the difference between the literary critic and geographer is that the former is ultimately concerned with the totality of the literary work, the geographer with his particular theme of study. Before turning to particular perspectives and types of literary engagement by geographers, further comments on the nature of literature and literary revelation may prove helpful.

The Nature of Literary Revelation

It is the deliberately cultivated subjectivity of the writer which makes literature literature and not, say, reporting. It is the work of the heart as well as the head; it is emotion, often recollected later, perhaps 'in tranquillity', when an earlier stimulus is reworked and given expression, in the manner perhaps that the painter in his studio may develop his quickly pencilled field sketch. Such a process is not to deny factual reproduction, for literature is often referred to as a mirror, reflection or microcosm of reality. Arnold Bennett, for example, is admired by Conrad for having achieved 'absolute realism'.[3] Again, a degree of isomorphism is no surprise when we know that authors visit locales of possible literary settings, or claim to need maps for the construction of stories no less than do generals for battles.[4] Such relationship is necessary for reasons of authenticity — otherwise we are in the realm of science fiction, fantasy or allegory — but it is not of the *essence* of literature.

Poetry, the novel, drama are distinguished by being the work of imagination rather than observation, creation rather than recording, fiction rather than fact. The best-known statement — or overstatement — expressing the supremacy of the creative faculty is Oscar Wilde's essay on 'The Decay of Lying',[5] referring to the decline in art occasioned by the rise in realism. But 'lying for the benefit of truth' is widespread among artists. H. E. Bates, for example, claims 'the business of writing fiction is an exercise in the art of telling lies . . . in making his readers believe his lies are truth but they are in fact truer than life itself'.[6]

The truth of fiction is a truth beyond mere facts. Fictive reality may transcend or contain more truth than the physical or everyday reality. And herein lies the paradox of literature. Although different in essence, and therefore a poor documentary source for material on places, people or organisations, literature yet possesses a peculiar superiority over the reporting of the social scientist. Ionesco, for example, in acknowledging the debt he owed to literature rather than textbooks for the 'living reality' of eighteenth-century France, writes that 'the genuine creator has a mysterious intuition of the concrete and particular truth which historians, sociologists and ideologies do not have'.[7] It is the same paradox which induces Flaubert to claim that 'Poetry is as precise as geometry. Induction is as accurate as deduction', adding that at that moment, without a doubt, his poor Bovary was suffering and weeping in twenty villages of France.[8] Literary truth has a universality: it evokes a response in Everyman's breast while apparently concerned with the particular. Again, it is a truth that is more humanly significant. That is why we may 'know' fictional characters better than those in real life, and why scholars acknowledge that literary authors are able to reveal more about human nature than are their fellow, specialist scholars.

> One learns much more from, for example, William Faulkner's great novel *Absolom, Absolom!* (on the guilt of slavery), from Tolstoi's *Anna Karenina* (on adultery), from Conrad's *Lord Jim* (on self-deception), from Dostoevsky's *Brothers Karamazov* (the problem of the use of power), than from the writings of many moral philosophers and most (if not indeed all) moral theologians.[9]

The writer here is a theologian—philosopher, and his object is not one of passing judgement on his colleagues — none, that is, that he would not apply to himself.

Associated with the universality of fictive reality is its inexhaustible character. Literary revelation, as opposed to reporting, is implicit, suggestive. This is so because the reader is no neutral receiver, but is also a creative, interpretative being. Both author and reader have their own unique biographical history and general social or cultural contexts, including the context of literary criticism, itself a creative influence.

The point of contact, the actual literary work, may be represented therefore as a small intersecting arc of the broad realms of artist and audience. The work is reactivated, and comes alive in the mind and feelings of the reader. A finite number of chapters or verses therefore has the power to create an infinite reality. Acknowledged by many authors, this quality of literary revelation is perhaps best summarised in Verlaine's terms of *vers donnés* and *vers calculés*.

The Geographer's Engagement with Literature

The *raison d'être* for the geographer's engagement with literature stems from the latter's above-mentioned universality. If T. S. Eliot could allow that his students were able to discover meanings in his works which had previously lain hidden to him as author, then it is possible that the geographer might find new insights in the *vers donnés* as he brings his own particular heritage into the realm of imaginative literature. Again, if philosophers, psychologists and others acknowledge the perspicuity of novelists and poets concerning the human condition, then, since the latter cannot be divorced from the environmental experience, literature would seem an appropriate field for the enquiring geographer. We are what we are largely as a result of our life's experiences, all of which have an integral environmental context or setting: events 'take place'.

The starting point of imaginative literature is the artist's gift of heightened perception and communication, which others acknowledge and profit thereby in different ways. Broadly, the geographer's engagement with literature in his study of place varies along a continuum between landscape depiction and human condition. On this continuum several distinctive types of engagement may be recognised.

Word painting is perhaps the most obvious overlap to spring to mind. 'Poets make the best topographers', claims Hoskins;[10] 'the skilful novelist often seems to come closest of all in capturing the full flavour of the environment', writes Meinig in Gilbertian sentiments of an earlier era.[11] We doubtless all have our favourite literary landscape depiction, where the quality of observation is more memorable and, indeed, more meaningful than the exactitude of conventional maps or tables of statistics for the same portion of the earth's surface. Geography perhaps approached closest to such holistic, 'flavoursome' description in the Vidalian studies of the personality of landscapes or regions; 'landscape signatures' has recently been suggested as a focus for renewed study combining geography and literature.[12]

The geography or topography *behind* literature was another early point of contact. Sir Archibald Geikie, for instance, at the beginning of this century turned his attention momentarily from the geology and glaciation of Britain to attempt to relate the influence of Britain's varied scenery to the type of literature produced.[13] A threefold division of the country into lowlands, uplands and highlands is followed by a discussion of the influence of such features as rushing torrents or placid streams, enclosed lowland or open moorland, the sea and so on. At a different scale, a less deterministic approach is best illustrated in the examination by Darby of Hardy's descriptions of South Wessex in relation to the five geological and physiographic subdivisions of Dorset.[14]

A more popular variant has been the search for geography *in* literature, treating writings as a literary quarry from which to construct a more general literary topography. Among geographers, for instance, Jay has made a detailed reconstruction of the Black Country from the works of Brett Young.[15] Such exercises, however, are more common among non-geographers, particularly by individuals or societies in the quest to uncover every association with their favourite literary son or daughter. The possible limited value to the geographer of such an approach is illustrated in the recent volume by Daiches and Flower.[16] Termed a 'narrative atlas', it adopts a 'map of the book' approach as it reconstructs the worlds of several British authors. In contrast stands the publication by Margaret Drabble, who goes over much the same ground but in order to elicit, and attempt to under-stand, different facets of human experience.[17] We are therefore partakers and not spectators. In summary, the contrast in approach is between critic and writer, scholar and artist.

Landscape depiction has a further, inherent, creative quality: it provides an important secondary source of knowledge, including environmental knowledge, and thus contributes to the general learning process whereby values, attitudes and aspirations are acquired, the end-product of which is our cultural refraction of reality.[18] Our cognitive frame of reference for viewing reality, therefore, may well be to varying degrees a *literary* frame of reference. The potency of literature as a creative force in this sense is seen in the way many parts of Britain are approached, 'seen' and remembered through the eyes of poets and novelists, whether we be native or visitor. Wilde, exaggerating to emphasise this formative element of art in the previously mentioned essay on lying, discusses how both nature and life imitate art, rather than vice versa. London's brown fogs and the quivering white sunlight

of France were, he claimed, the invention of the Impressionist school of
painting, while in literature

> The world has become sad because a puppet [Hamlet] was once
> melancholy. The Nihilist . . . was invented by Tourguenieff, and
> completed by Dostoevski. Robespierre came out of the pages of
> Rousseau as surely as the People's Palace rose out of the *débris* of a
> novel . . . The nineteenth century, as we know it, is largely an
> invention of Balzac.[19]

The list could be easily extended – Christmas being the invention of
Dickens or Scottish history that of Sir Walter Scott, for instance – for
the power of suggestion is very pervasive. The manner in which family
likenesses are recognised most surely by those least acquainted with the
persons, or the way personalities come to resemble the cartoonist's
sketch, and even pet-owners their pets, all bear witness to this
everyday phenomenon.

The contemporary influence of imaginative literature as a source of
environmental knowledge might be questioned in view of the small part
it constitutes of the vast, heterogeneous output in the field of publications, quite apart from the impact of more recent communication
media. Creative literature, therefore, is hardly a moulder of morals as it
was up to the time of Hardy's Tess, and gone are the days when an
eminent politician could consider novel-writing 'the best chance of
influencing opinion'.[20] Priestley, in his review, *Literature and Western
Man*, considers modern literature to be 'off-centre' and introverted with
little to say to modern man,[21] while Steiner has written generally on
the retreat of literary culture.[22] Now, while not denying the widespread
influence of visual media and the 'popular' arts, it should be noted that
many classic works are themselves receiving wider dissemination than
ever before through projection on the television and cinema screens.
Further, one should not underestimate the formative influence of
formal education, which for generations has inculcated particular
views and values. Literature as a component part of this process has
therefore been, and remains, an agency of 'social control'.[23] It may be
further noted that it is from those who have undergone the greatest
exposure to this type of learning process that the decision-makers in the
various walks of life have traditionally been drawn. Thus, to give one
example, it has been suggested that the consistent negative portrayal, or
plain neglect, of the North of England at the expense of the South in
novels and poems, has contributed to a Southern-based or -biased

perceptual frame of reference among the political, financial and business decision-makers in Britain. The strength and potency of this communicated image is reflected in the recent efforts of regional agencies to propagate a counter-image.[24]

It will be evident from earlier discussion that the projection of what we term 'false' geographies is in fact not so exceptional but, rather, central to the very nature of literary revelation. The portrayal of landscapes as they ought to be, or could be in some idyllic way, can be traced back through literature to classical times. This so-called 'pastoral convention' − with its converse, the 'counter-pastoral' − has been extensively examined by Marx and Williams.[25] Literary refraction therefore creates the basis for a *new* perception, as much as providing the basis for a deeper, 'cleansed' perception.

The deepest engagement with imaginative literature, concerned most fully with both internal and external phenomena, comes from geographers exploring the nature and aspects of environmental experience as part of the human condition. The starting point is acknowledgement of the artist's perceptive insight: literature is the product of perception, or, more simply, *is* perception. The writer therefore articulates our own inarticulations about place, our fellow men and about ourselves, providing thereby a basis for a new awareness, a new consciousness. New windows are opened, but, perhaps more important, old ones also, so that, in Natanson's words, 'the experiential foundation of our world' is revealed: 'the horizon of daily life and the a prioris which attend it . . . What we ordinarily take for granted . . . is rendered explicit.'[26] In the last analysis, therefore, it is perhaps not so much what the poet, novelist or playwright says, but what he *does* to us that matters, in both enlarging our sensitivity and our capacity for subsequent experience.

Literature, then, is both a source for new insights and a testing ground for hypotheses in exploring 'the experiential foundation of our world'. Key concepts to emerge are those which focus on insideness−outsideness,[27] our 'lived reciprocity',[28] or the dialectic between rest and movement.[29] At-homeness and rootedness on one hand, exile and restlessness on the other, are important polar foci for study in a society increasingly mobile and a world of increasing homogenisation.

The role of physical, earthy objects in our environmental experience also warrants study. What is it, for instance, that makes Sillitoe, a novelist of entirely urban background, write that 'real love begins with one's feeling for earth, and that if you do not have this love, you cannot really begin to love people'?[30] Or Ransome, with an upbringing in

Leeds, to dip his hand into Coniston on his annual visit to the Lake District 'as a proof to myself that I had indeed come home'?[31] Returning exiles — soldier, statesman, monarch — kiss the earth on which they rejoice to set foot again. Television permits millions to observe the spiritual leader of the Catholic church greet the earth of new lands and continents with the same obeisant gesture, symbolic that man himself is made from the dust of the ground.

Although we believe it is the social dimension which gives life to place, how many share the experience of Olsson, who, with time, finds human touch fading away:

> What I remember better than I know is the roundness of the stone in my hand before I throw it and the sharpness of the grass in which I was hiding from the chasers. Not people but stones and grass. Not the shifting nature of humans but the stable humaneness of nature.[32]

This experience with intimacies and seeming trivia is part of the same continuum whereby the memory indelibly records major events by fusing place and person in what Hardy termed 'permanent impressions'. (Alternative terms — 'epiphanies' of Joyce and 'privileged moments' of Natanson — are perhaps more meaningful appellations for such experience.)

The stability of the physical — 'the stable humaneness of nature' — anchors the fleeting seconds of the successive presents as we are transported between past and future, suspended permanently between our memories and our hopes. At heart, our experience is personal, our memories and our pronouncements self-referential. And there's the irony: accusation of mere self-reflection by some of our colleagues of alternative philosophic persuasion is willingly conceded by geographers immersed in this last type of literary engagement which is seen as the most relevant task, not peripheral, not decorative, but central.[33] Cognitive development theorists may proclaim progression beyond an initial egocentric stage, but the poet cries for a viewpoint outside himself and the religious seek a saviour who will save them — from themselves. We are alone, then, but no more so than the artist, to whose ranks we belong and whose utterances we value on our search for insight into our condition.

Essays on the Experience of Place

The interplay between geography and literature summarised above is illustrated in the following collection of essays. Their unifying characteristic is that all the authors have been socialised within the discipline of geography and are here eschewing a positivistic framework for alternative humanistic perspectives as they explore imaginative literature for insights into the experience of the man–environment relationship. (For a discussion of the characteristics and variety of humanistic approaches, the reader is referred to a companion volume by Ley and Samuels.[34]) The concept of place provides an organising concept for what is termed our immersion in, or interpenetration with, the world. With its experiential perspective and varied scale, place relates to an area which is bounded and has distinctive internal structure, to which meaning is attributed and which evokes an affective response.[35] The natural result of familiarity with place, however, may require a stranger's questioning or actual severance in order to articulate a response which in any taken-for-granted world lies dormant. Physical place is 're-placed' through our sensibilities by an image of place, which is no less real, while the phenomenon of sense or spirit of place highlights the experiential nature of our engagement. In terms of scale, place may refer to one's favourite chair, a room or building, increasing to one's country or even continent.

The essays fall into three distinct, but by no means mutually exclusive, groupings. Since literature is our source material, the first contributions explore aspects of the nature of literature in relation to experience of place. Thus, Ruskin's three orders of truth in art and literature are examined in relation to his work on clouds. The autonomous role of literature is then discussed with reference to several, diverse examples. Being 'estranged' from reality, literature is shown to present a different reality. The humanist and radical, however, put different interpretations on the relationship or interplay, the humanist viewing literature as an individual's insight and reflection on the world, while the radical view is more socially reflective. In putting more emphasis on the external social processes affecting the artist, the latter's interest is on the value system held and the consequent type of 'consciousness' promoted.

A second grouping of essays is concerned with particular conditions of environmental experience. The significance of early place is shown in studies of the newcomer's transformation from conditions of outsideness to insideness, and through an examination of the phenomenological notions of at-homeness and rootedness. A further, unorthodox essay, however, emphasises that the truth of yearning for home lies not in things or persons yearned for, but in the very process of yearning itself. And here the challenge is in trying to understand and capture a double dialectic of symbol and meaning.

Contributions to the third grouping adopt a strong regional theme while illustrating particular aspects of the experience of place. Literature as an agent of 'social control' is discussed in its influence on fashioning the early tourism of the Lake District and in the Mediterranean, while study of the Joad family's westward trek across America to California explores the didactic value of literature, showing, in this instance, its suitability as a take-off point in cultural geography. Literature on nineteenth-century Boston reveals early evidence of the middle-class ambivalence to the modern city, while nineteenth-century St Petersburg provides an interesting example of the dialectic between writer and place whereby physical reality is transcended in the minds of those who experience it. The Suffolk of Crabbe's poetry provides further evidence of the same process, particularly how individuals' moods are echoed and expressed in the landscape about them, while the Shropshire of Mary Webb provides the setting for a discussion of reality and symbol.

Any collection of this kind lends itself either to a selective or a continuous engagement by the reader. Whichever is chosen, the value of the contributions is to be judged in relation to the reader's own environmental experience. Confirmation and enlightenment, it is hoped, will be among the reactions to the following essays.

Notes

1. Yi-Fu Tuan, 'Literature and Geography: Implications for Geographical Research' in D. Ley and M. S. Samuels (eds.), *Humanistic Geography: Prospects and Problems* (Croom Helm, London, 1978), pp. 194–206.

2. C. S. Lewis, *Selected Literary Essays* (Cambridge University Press, Cambridge, 1969), pp. 301–11.

3. But see Virginia Woolf, *Mr. Bennett and Mrs. Brown* (Hogarth Press, London, 1924).

4. A. Sillitoe, *Mountains and Caverns* (Allen & Unwin, London, 1975), p. 68.

5. Oscar Wilde, 'The Decay of Lying' in R. Ellmann, *The Artist as Critic: Critical Writings of Oscar Wilde* (Allen & Unwin, London, 1970), pp. 290–320.

6. H. E. Bates, *The Vanished World: An Autobiography* (Michael Joseph, London, 1969), vol. 1, p. 151.

7. E. Ionesco, cited in A. de Jonge, *Dostoevsky and the Age of Intensity* (Secker & Warburg, London, 1975), p. 1.

8. G. Flaubert, 'Letter to Louise Colet, 1853', cited in P. Stevick, *The Theory of the Novel* (Free Press, New York, 1967), p. 391.

9. D. MacKinnon, *Explorations in Theology – 5* (SCM Press, London, 1979), pp. 74–5.

10. W. G. Hoskins, *The Making of the English Landscape* (Penguin, Harmondsworth, 1970), p. 17.

11. D. W. Meinig, 'Environmental Appreciation: Localities as a Humane Art', *The Western Humanities Review*, vol. 25 (1971), p. 4.

12. C. L. Salter and W. J. Lloyd, 'Landscape in Literature', *Resource Papers for College Geography* (Association of American Geographers, Washington DC), no. 76–3 (1977).

13. Sir Archibald Geikie, *Landscape in History and Other Essays* (Macmillan, London, 1905), pp. 76–129.

14. H. C. Darby, 'The Regional Geography of Thomas Hardy's Wessex', *Geographical Review*, vol. 38 (1948), pp. 426–43.

15. L. J. Jay, 'The Black Country of Francis Brett Young', *Transactions of the Institute of British Geographers*, vol. 66 (1975), pp. 57–72.

16. D. Daiches and J. Flower, *Literary Landscapes of the British Isles* (Paddington Press, London, 1979).

17. Margaret Drabble, *A Writer's Britain: Landscape in Literature* (Thames & Hudson, London, 1979).

18. Yi-Fu Tuan, 'Literature, Experience and Environmental Knowing' in G. T. Moore and R. G. Golledge (eds.), *Environmental Knowing: Theories, Research and Methods* (Dowden, Hutchinson & Ross, Stroudsburg, Pa., 1976), pp. 260–72.

19. Wilde, 'The Decay of Lying' in Ellmann, pp. 308–9.

20. B. Disraeli, cited in K. Tillotson, *Novelists of the 1940s* (Oxford University Press, London, 1954), p. 120.

21. J. B. Priestley, *Literature and Western Man* (Heinemann, London, 1960), p. 444.

22. G. Steiner, *Language and Silence: Essays 1958–1966* (Faber & Faber, London, 1967).

23. J. Rockwell, *Fact in Fiction: The Use of Literature in the Systematic Study of Society* (Routledge & Kegan Paul, London, 1974), pp. 3–4.

24. D. Pocock and R. Hudson, *Images of the Urban Environment* (Macmillan, London, 1978), pp. 122–7.

25. L. Marx, *The Machine in the Garden* (Oxford University Press, London, 1964); R. Williams, *The Country and the City* (Oxford University Press, London, 1973).

26. M. Natanson, *Literature, Philosophy and the Social Sciences* (Martinus Nijhoff, The Hague, 1962), p. 97.

27. E. C. Relph, *Place and Placelessness* (Pion, London, 1976).

28. A. Buttimer, 'Home, Reach and the Sense of Place' in A. Buttimer and D. Seamon (eds.), *The Human Experience of Space and Place* (Croom Helm, London, 1980).

29. D. Seamon, *A Geography of the Lifeworld* (Croom Helm, London, 1979).

30. Sillitoe, *Mountains*, p. 73.

31. Arthur Ransome, cited in Drabble, *A Writer's Britain*, p. 262.

32. G. Olsson, Chapter 7 of this volume.

33. D. Lowenthal and H. C. Prince, 'Transcendental Experience' in S. Wapner, S. B. Cohen and B. Kaplan (eds.), *Experiencing the Environment* (Plenum, New York, 1976), p. 118.

34. Ley and Samuels, *Humanistic Geography*.

35. Yi-Fu Tuan, *Space and Place: The Perspective of Experience* (University of Minnesota Press, Minneapolis, 1977).

2 OF TRUTH OF CLOUDS: JOHN RUSKIN AND THE MORAL ORDER IN LANDSCAPE

Denis Cosgrove and John E. Thornes

> . . . up above, what wind-walks! what lovely behaviour of
> silk-sack clouds! has wilder, wilful wavier meal-drift moulded
> ever and melted across the skies?
>
> Gerard Manley Hopkins

The sky and its clouds condition much of our experience of place, they give the mood of landscape. Although frequently ignored in geographical writing, the imaginative writer finds in them a source of powerful metaphor for relating human feeling to the physical milieu.[1] For the landscape artists they are essential, for it is in sky and clouds that the source of light for their painting is to be found. For this reason John Ruskin, the most geographical of commentators on, and critics of, landscape painting,[2] devoted a considerable proportion of his writing to understanding the form, organisation and meaning of clouds. His efforts were devoted to three aspects of cloud study: the scientific description and understanding of their form and structure; the analysis of that beauty which derived from the involvement of the human subject with clouds as an object of landscape study; and the moral questions they raised concerning the authenticity of human existence as it is expressed in the experience of landscape and the creation of places. For Ruskin, the purpose of landscape art was to 'serve' nature, and his discussion of clouds parallels in a very direct way the manner in which geographers have employed landscape painting in the service of their own ends of understanding man and environment.

Geography and Landscape Painting

Truth: Landscape as a Factual Source

A geographical understanding of landscape rests primarily upon the accurate description of its facts, forms and relations. To this end landscape painting may provide a documentary source, particularly for the study of past landscapes. We are aware of the employment of topographical draughtsmen for the purposes of recording military or naval surveys before the invention of photography. Similarly, English amateur artists of the eighteenth-century gentry made use of water-colours to record the landscapes of their 'Grand Tours' rather as the modern tourist makes use of his colour slides. For academic painting, the influence of this tradition upon the acceptance of landscape as a subject, and watercolour as a medium, was considerable.[3] The attention to the accurate observation of form for landscape painting paid by this tradition bears fruit in the continuing value of Constable's cloud study to meteorologists.[4] But even paintings from an era when idealised schemata rather than direct reconstruction determined the representation of landscape have been used as documentary evidence for the geographical study of historical landscapes.[5]

There are limitations to the use of painting as a source for the accurate reconstruction of past landscapes. Techniques of painting change, as do artistic conventions. Perspective rules, frame shape and medium each controls the degree of 'accuracy'. For example, the rapid execution of actual scenes as finished pictures depended upon the availability of paints which were easily portable out of doors in a range of weather conditions. Again, for a long time, convention required that a landscape be 'composed' of elements chosen from a series of sketches according to a scheme which excluded inappropriate forms and added others in order to achieve an idealised landscape.[6] The 'landscape of fact' is always mediated through a set of rules, limitations and individual intentions on the part of the painter, of which the geographer employing his work as a documentary source is not always fully aware.

Beauty: Landscape as a Concept Unifying Man and Nature

Both geographers and historians of art have noted that the word 'landscape' conveys a dual sense: of a delimited area, and of a visual unit – an object which engages the subject in a pictorial sense. Commenting upon the poet Gerard Manley Hopkins' concept of 'inscape', Peters says of its suffix, 'scape'

> posits the presence of a unifying principle which enables us to consider part of the countryside or sea as a unit and as an individual, but so that this part is perceived to carry the typical properties of the actually undivided whole.[7]

This holistic quality of landscape is explicit in painting. The frame forces a unity so that even the painting of a landscape untouched by evidence of human agency requires a subject – the viewer. The subjective involvement in landscape implied in the word itself has troubled many who regard geography as a positive science,[8] but others have seen in it the fountainhead of the 'geographical imagination'. Even Carl Sauer's enunciation of geography as a positive science recognised that 'a good deal of the meaning of area lies beyond scientific regimentation . . . Having observed widely and charted diligently there yet remains a quality of understanding at a higher plane that cannot be reduced to formal processes.'[9] For some geographers, it is in landscape painting that this quality is best expressed, so that 'the history of art . . . provides the largest corpus of instructive literature available to the student of the cultural land-scape'.[10] When geographers themselves have turned to artistic techniques to capture the unity of subject and object in landscape they have generally made poor artists and poets, but if they are prepared to learn the academic skills of the historian of art or literature there is much that the geographer can bring to the understanding of human subjectivity in landscape, through its imaginative expression in painting, poetry and prose.

Relation: Landscape as a Moral Issue

There is a third dimension of the geographical concern with landscape which goes beyond its accurate description and its subjective unity. This is the moral dimension. Popular discontent with the contemporary treatment of our environment finds its academic expression in a revived humanism and radicalism in geography. This has focused attention on issues concerning the meaning of place and landscape and the conditions under which it emerges and is expressed. Many suspect that geography's adoption of scientific method to the exclusion of alternative approaches, and its alliance with corporate planning and environmental control techniques, lock the discipline institutionally and ideologically into an alienating social system and make geography partly responsible for the creation and legitimisation of inhuman, placeless landscapes for an inauthentic human existence. Both humanistic and radical geography are grounded in a moral conviction that geography must be reflexive in its treatment of landscape, that the object of geographical study requires the commitment of its subject. Such moral concerns refuse to accept the exclusion of values from geographical study.[11]

Landscape painting is itself one aspect of the elaborate cultural structure of historically specific societies. It offers an area for humanist and radical geographers to explore relationships between environment attitudes and human values under varying forms of social appropriation of nature. It tells of meaning given to landscapes under specific historical conditions and, as Panofsky pointed out, it is in the search for meaning that the humanities meet on a common place, rather than serving as handmaidens to each other.[12] Geography, from its profound interest in the world as the home of man, has much to contribute to our understanding of human expression of meaning in landscape and to the moral debate which inevitably conditions both meaning and the possibilities of its expression.

In all three dimensions of geographical interest in landscape and landscape painting, John Ruskin had already opened up questions a century ago. In his treatment of clouds many of these questions are directly raised, for, like Hopkins, Ruskin looked 'up at the skies' for more than one level of understanding.

Ruskin's Landscape Study

If there is a unity in Ruskin's seemingly divergent writings on art and on social and political questions, it springs from his conviction that the earth was created, and required maintenance, as a fitting place for a true humanity. The truth of that conviction was, for him, made manifest in the study of landscape:

> You cannot have a landscape by Turner, without a country for him to paint, you cannot have a portrait by Titian without a man to be portrayed . . . I can get no soul to believe that the beginning of art is in getting our country clean, and our people beautiful.[13]

However, each of the three dimensions of landscape study discussed above was given encouragement by Ruskin's education and early life in suburban South London of the 1820s and 1830s. His natural talent for drawing was trained and disciplined by a series of drawing masters. Among them were Copley Fielding, J. D. Harding and Samuel Prout, early members of the Society of Painters in Water Colours, an institution founded to establish the use of watercolour as an acceptable medium for academic art and to gain recognition also for landscape as a valid subject for artistic treatment in painting. Particular stress was placed by these artists and teachers upon the accurate observation and representation of landscape forms, thereby providing Ruskin with a foundation not only for his own detailed drawing of clouds, rocks, plants and animal life, but also for his belief in the canons of scientific accuracy in artistic observation.

Ruskin was born in 1819, at the height of Romantic interest in nature, and he was nurtured upon the Waverley novels of Walter Scott and his family's travels through the Romantic regions of Britain and Alpine Europe. The sense of harmony between man and wild nature which he was to explore in so much of his writing does not belong to Ruskin alone, it was part of the intellectual inheritance of his period and his class. But for Ruskin it was necessary, during a life which spanned the remainder of the nineteenth century, to reformulate his Romantic inheritance in the face of extraordinary changes, not only in the physical environment itself and man's power of intervention in it, but also in the human understanding of the processes of the natural world and of men's social organisation.

The moral cogency of landscape was elaborated by Ruskin in the application to it of a form of biblical exegesis, learned from his mother. Landscape, like the Bible, he held to contain both a visible, surface truth apparent to the disciplined observer, and a deeper 'beauty'. It was the mind which could apprehend this hidden message whereby God was revealed in his purity, goodness, truth and mercy to men. This 'geoteleology' required that the human subject treat landscape beauty as the record of the divine, and through a dialectic of eye and mind discover the moral purpose of landscape. The duty of art was to declare this purpose and thus inform human relations and human endeavour.

For Ruskin then, landscape was an object of scientific observation of 'truth', an embodiment of divine revelation in its 'beauty', and a source of moral instruction in its 'relations'. The three dimensions are inseparable, so that the imaginative treatment of landscape in art is firmly grounded in its object, but equally has a didactic purpose for its subject. As Ruskin declared to his Oxford students in 1871:

> Landscape painting is the thoughtful and passionate representation of the physical conditions appointed for human existence. It imitates the aspects, and records the phenomena, of the visible things which are dangerous or beneficial to men; and displays the human methods of dealing with these, and of enjoying them or suffering from them, which are either exemplary or deserving of sympathetic contemplation.[14]

In the study of clouds this articulated unity of the dimensions of landscape interest is clearly demonstrated.

Meteorological Truth

Ruskin was interested in meteorology from an early age. When only seven he adapted a passage from Jeremiah Joyce's *Scientific Dialogues* about a thunderstorm:

> Harry ran for an electrical apparatus which his father had given him and the cloud electrified his apparatus positively after that another cloud came which electrified his apparatus negatively . . . Harry began to wonder how electricity could get where there was so much water . . . [15]

The final thought was Ruskin's own — his mind was already puzzling over atmospheric phenomena.

When he was sixteen Ruskin kept a weather diary of a journey abroad with his parents; he also constructed his own early cyanometer (an instrument to measure the blueness of the sky) based on that invented by De Saussure.[16] As a youthful member of the Meteorological Society of London, at the age of eighteen, he delivered his first paper entitled 'On the Formation and Colour of such Clouds as are caused by the agency of Mountains'. This paper was not published, but two years later in 1839 he published a remarkable paper entitled 'Remarks on the Present State of Meteorological Science', in which he spelled out the necessity of worldwide weather observations for dynamic meteorology:

> There is one point, it must now be observed, in which the science of meteorology differs from all others. A Galileo, or a Newton, by the unassisted workings of his solitary mind, may discover the secrets of the heavens, and form a new system of astronomy. A Davy in his lonely meditations on the crags of Cornwall, or in his solitary laboratory, might discover the most sublime mysteries of nature, and trace out the most intricate combinations of her elements. But the meteorologist is impotent if alone; his observations are useless; for they are made upon a point, while the speculations to be derived from them must be on space . . . The Meteorological Society, therefore, has been formed, not for a city, nor for a kingdom, but for the world.[17]

This is certainly true for studies of the general circulation of the atmosphere, and Ruskin rightly steered clear in his writings from dynamic meteorology. Instead he devoted his energies to the analysis and description of more obvious local weather phenomena — aspects of physical meteorology. An individual *can* speculate upon the origins of a cloud, or the blueness of the sky, without the need for global observations.

Ruskin's knowledge of meteorology in relation to the newly developing science of meteorology in the nineteenth century accelerates and declines in fits and starts. It can be traced through examining the contents of *Modern Painters*, in which there are two large sections on atmospheric phenomena: Volume I, part 2, entitled 'Of Truth', published in 1843, and Volume V, part 7, published seventeen years later.

Modern Painters I

Most of *Modern Painters I* is concerned with empirical accuracy. The section entitled 'Of Truth of Skies' is split into four chapters — of the open sky, of the region of the cirrus, of the central cloud region and of the region of the rain-cloud. In a footnote to the first and second editions Ruskin states:

> I shall often be obliged, in the present portion of the work, to enter somewhat tediously into the examination of the physical causes of phenomena, in order that in the future, when speaking of the beautiful, I may not be obliged to run every now and then into physics . . . I must be allowed . . . to spend sometimes almost more time on the investigation of nature than on the criticism of art.[18]

Ruskin's knowledge of meteorology at the time of writing *Modern Painters I* was reasonably up to date,[19] but meteorology was still an infant science. For instance, in attempting to explain the colour of the sky Ruskin states: 'This is of course the colour of the pure atmospheric air . . . the total colour of the whole mass of that air between us and the void of space.'[20] Ruskin cannot be severely criticised for his lack of truth, for Lord Rayleigh's famous paper explaining the scattering of light by the atmosphere did not appear until 1871.[21] In a later work, *The Queen of the Air*, published in 1869, Ruskin acknowledges this new meteorological theory.[22]

In the chapter 'Of Truth of Skies' Ruskin explains concisely why under a cloudless sky the air is often hazy. 'Aqueous vapour or mist, suspended in the atmosphere becomes visible exactly as dust does in the air of a room — so that a transverse sunbeam is a real obstacle to vision, you cannot see things clearly through it.'[23] Ruskin's observations

on light and colour in the atmosphere are always astute,[24] although
physical explanations are not always as simple as in the last example.
Accurate descriptions of physical phenomena are very important in
meteorology, not only to stimulate explanations of hitherto unnoticed
effects, but also because it is often difficult and expensive to make
measurements of atmospheric phenomena above the earth's surface.

The three chapters on clouds show that Ruskin was familiar with
Luke Howard's cloud classification which was first published in 1803.[25]
However, Ruskin's division — 'the upper region, or region of the cirrus;
the central region, or region of the stratus; the lower region, or the
region of the rain-cloud'[26] — is interesting in concept but confusing in
content. The association of the upper region with cirrus is sensible, but
the central region should not only be associated with stratus which
was defined by Howard to be 'the lowest of the clouds, since its
inferior surface commonly rests on the earth or water'.[27] Ruskin
probably erred unwittingly, having correctly associated his central cloud
region with cumulus in his chapters on cloud beauty in Volume V of
Modern Painters. Indeed most of the chapter on the central cloud
region in Volume I relates to cumulus! The third division — that of the
rain-cloud — is confusing in that rain-clouds can extend to great heights
in the atmosphere, especially in thunderstorms. The concept of three
regions of cloud is interesting, nevertheless, since the idea of distin-
guishing clouds on the basis of their height was not generally recognised
until Perou proposed the idea in 1855.[28] Howard's classification is
based on cloud shape, rather than cloud height, although Howard's
three main cloud types — cirrus, cumulus and stratus — are suggestive
of three layers.[29]

At the time Ruskin wrote *Modern Painters* meteorologists did not
realise that cirrus clouds are composed of ice crystals. Ruskin defined
the lower limit of his upper cloud region to be 15,000 feet. We now
know that altostratus and altocumulus (which are not composed of ice
crystals, but usually supercooled water droplets) can be found above
that level. It is not clear, however, whether or not Ruskin labelled
those clouds as cirrus. On the one hand, he points to a multitude and
order of arrangement of cirrus that is more typical of altocumulus, but
on the other hand, he does write of 'masses like flocks of sheep: such
clouds are three or four thousand feet below legitimate cirrus',[30] which
suggests that he could distinguish the lower altocumulus. Nowhere does
Ruskin mention cirrostratus or cirrocumulus to clarify the situation.
Hence his definition of cirrus is confusing to a modern reader, but no
more confusing perhaps than Howard's original classification.

The chapter on the central cloud region also requires careful scrutiny. Ruskin states that the clouds in this region may 'be considered as occupying a space of air . . . extending from five to fifteen thousand feet above the sea',[31] which seems sensible. He then confuses the issue by writing: 'Even in nature, these clouds are comparatively uninteresting, scarcely worth raising our heads to look at . . . yet they are, perhaps, beyond all others the favourite clouds of the Dutch masters.'[32] Ruskin does not mean cumulus, the cloud he continuously lavishes with praise, but presumably stratocumulus, a term coined by the German meteorologist Kaemtz in 1836. However, stratocumulus is usually a low-level cloud, and its base in Britain is more often than not below 5,000 feet. Ruskin was probably over-influenced by cloud formation in the Alps, which may be confusing for the general reader. There is a short section on the formation of helm clouds — stationary clouds on mountain tops — but Ruskin's theory on their formation is wrong, as he readily admits in *Modern Painters*, Volume V.

It is when he describes cumulus clouds that Ruskin's prose begins to flow with most power:

We are little apt, in watching the changes of a mountainous range of cloud, to reflect that the masses of vapour which compose it, are huger and higher than any mountain range of the earth; and the distances between mass and mass are not yards of air traversed in an instant by the flying form, but valleys of changing atmosphere leagues over; that the slow motion of ascending curves, which we can scarcely trace, is a boiling energy of exulting vapour rushing into the heaven a thousand feet a minute; and that the toppling angle whose sharp edge almost escapes notice in the multitudinous forms around it, is a nodding precipice of storms 3,000 feet from base to summit. It is not until we have actually compared the forms of the sky with the hill ranges of the earth, and seen the soaring Alp overtopped and buried in one surge of the sky, that we begin to conceive or appreciate the colossal scale of the phenomena of the latter . . . every boiling heap of illuminated mist in the nearer sky, is an enormous mountain, fifteen or twenty thousand feet in height, six or seven miles over an illuminated surface, furrowed by a thousand colossal ravines, torn by local tempests into peaks and promontories, and changing its features with the majestic velocity of the volcano.[33]

Hough argues that with these vivid descriptions Ruskin was inventing a new line of research of his own:

> It is to the re-creation of the sense of sight as applied to nature that most of *Modern Painters* is devoted: hence the long sections on truth of clouds, on truth of rocks . . . in fact he is inventing a new line of research of his own, different in aim from scientific study of natural forms, and different too from the eighteenth-century study of the picturesque. His description of . . . clouds . . . tells us nothing about the weather or the way such clouds are formed . . . but it reveals a beauty and complexity in their organization that nine out of ten ordinary observers would never have been able to see for themselves.[34]

Lehrs takes this a stage further:

> By virtue of his pictorial-dynamic way of regarding nature, Ruskin was quite clear that the scientists' one-sided seeking after external forces and the mathematically calculable interplay between them can never lead to a comprehension of life in nature. For in such a search man loses sight of the real signature of *life*: form as a dynamic element. Accordingly, in his *Ethics of the Dust*, Ruskin does not answer the question: 'What is Life?' with a scientific explanation, but with the laconic injunction: 'Always stand by Form against Force.' This he later enlarges pictorially in the words: 'Discern the moulding hand of the potter commanding the clay from the merely beating foot as it turns the wheel.'[35]

Thus Ruskin's scientific understanding might be said to be the opposite of that of Constable as expressed in the latter's famous adage, 'we see nothing truly till we understand it'.[36]

The chapter on the rain-cloud finds Ruskin using evidence from his Alpine experiences. He notes that snow is found on the top of the highest peaks and assumes that this falls from either cirri or clouds from the central region. He states that he has never been in a violent storm at a height above 9,000 feet. He is therefore not aware at this stage of his life of the different types of rain-cloud. A thunderstorm (cumulonimbus) can extend from 1,000 feet up to the tropopause at an altitude twice that of the highest Alps; similarly, the stratus or nimbo-stratus of a frontal system. The close proximity of these cloud types to the ground obviously obscures their magnitude. Ruskin stares at their bases and gives an interesting description of 'mamma' cloud, for instance, when discussing the cloud painting of Copley Fielding.[37]

Modern Painters V

In the seventeen years that elapsed between the publication of Volumes I and V of *Modern Painters* Ruskin's desire to understand the basic processes of meteorology continued. Despite the continuing attention to cloud form in Volume V, Ruskin was obsessed with cloud physics. The more he thought about clouds, the more Ruskin realised how little he knew, and indeed how little was generally known, about their formation and lifespan. Writing in 1872 in *Eagle's Nest*, he states: 'I was obliged to leave *Modern Painters* incomplete, or, rather, as a mere sketch of intention, in analysis of the forms of cloud and wave, because I had not scientific data enough to appeal to.'[38]

Thus the chapters on cloud beauty are filled with questions as regards cloud physics. The first chapter, entitled 'The Cloud-Balancings', is an attempt to set down all that he knew of cloud microphysics, and to raise questions about those things that he did not understand. His chief problem was that of how a cloud can apparently 'float' in the atmosphere when water is approximately eight hundred times denser than air. This mystery Ruskin thought perhaps to be beyond man's grasp:

'Knowest thou the balancing of the clouds?' Is the answer ever to be one of pride? 'The wondrous works of Him which is *perfect* in knowledge.' Is *our* knowledge ever to be so?[39]

He then admits that he has not kept up his meteorological studies, but attempts to explain why clouds 'float' using what was called by meteorologists the vesicular hypothesis. This hypothesis supposed that clouds are composed of bubbles of water, and the height at which a bubble floats is determined by the size of its evacuated hollow interior. The survival of the vesicular hypothesis was an embarrassment to meteorological science. Despite the fact that no one could explain the formation of vesicles the theory was not effectively demolished until 1871.[40] We cannot blame Ruskin therefore for advancing this theory, although he felt that the biblical title of the chapter — 'Cloud-Balancings' — supported the theory. He did qualify his thought about the theory, it being 'a possibility only, not seeing how any known operation of physical law could explain the formation of such

molecules'.[41] In 1884 Ruskin received a letter from Professor Lodge of
Liverpool University which is included as a postscript to the chapter on
'The Cloud-Balancings' in the *Works*. In the letter Lodge explains
briefly why vesicles of water cannot exist. In the following year Ruskin
wrote to Miss Kate Greenaway: 'This has been a very bright day to me,
I've found out why clouds float, for one thing!!! and think what a big
thing that is.'[42]

Thus quite late in Ruskin's career he at last caught up with the
advancing meteorological theories, and indeed he intended to publish a
series entitled *Coeli Enarrant − studies of cloud form and of its visible
causes*. Unfortunately, only the first part was published, in 1885,[43]
although proofs for the second part were prepared which contained the
postscript from Professor Lodge.

The next chapter, entitled 'The Cloud-Flocks', was, with minor
alterations, to have been the second part of *Coeli Enarrant*. The
alterations that Ruskin proposed, given in the *Works* volume, are few
in number, since it was his intention to convey the beauty of cloud
form in this and the next chapter. The 'Cloud-Flocks' are the clouds of
Ruskin's upper region as defined in *Modern Painters I*. His power of
observation is illustrated by an occasion one winter morning when he
recorded approximately 50,000 separate cloud segments arranged in
ranks across the sky. He questions how they could have been formed,
but does not attempt an explanation, believing it to be something to do
with atmospheric electricity (in meteorological vogue at the time).

The next section deals with cloud perspective in the sky (see Figures
2.1 and 2.2). There had been nothing like it in meteorological literature,
and yet Ruskin was going to withdraw the section on cloud perspective
from the *Coeli Enarrant* edition because he had 'never heard any one
express the slightest interest, or intimate that they had put them to any
use'.[44] Damisch criticises Ruskin's analysis of perspective in that it
failed to break away from conventional pictorial perspective, but
rather forces nature to submit to it.[45] Running through all Ruskin's
analysis of form, not just of clouds, is the consistent curving line
expressed in the bird's wing, the leaf shape, the scree slope or the
cirrus cloud. In order to portray this curving line in cloud form, Ruskin
presented a rigid cube perspective − an artistic convention that defined
the limits of accuracy for him. The diagrams represent an increasing
complexity of perspective drawing from the point of a single observer
standing at ground level, and fitted into a rectangle − the conventional
frame shape for a picture.

Turner, in contrast, made a concentrated effort to escape such a

Figure 2.1: John Ruskin's Cloud Perspective (Rectilinear)

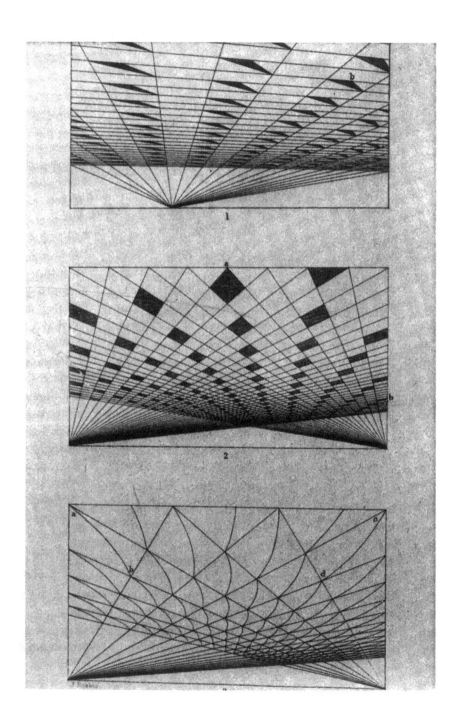

Figure 2.2: John Ruskin's Cloud Perspective (Curvilinear)

point of view and such a rigid perspective — for example, by strapping himself to the mast of a ship in a storm to get into the picture; disposing of the 'ground' in the Petworth interiors; altering frame shape etc. Such efforts were in order to develop his vortex systems, thus representing not a conventional 'visual' truth to the external observer, but another inner truth of the normally invisible eddies of atmospheric motion. Ruskin faced with this vortex has little to say, but perhaps this is one of the reasons why he wanted to discard his perspective drawings.

The third chapter in this section is called 'The Cloud-Chariots' and describes the cumulus. It was not revised by Ruskin for *Coeli Enarrant*, although no doubt it would have been if time had permitted. He gives what he calls a plausible explanation of cumulus formation based on the uplift of warmed air, but then asks how the cloud is held together in such 'heaps'. He could understand the flat bases of cumulus clouds — at the condensation level — but he could not understand the equally distinct edges of the cloud. The physics of the growth and paracme of cloud droplets was little understood at the time by meteorologists, and even today the shape of clouds is rarely examined.

Ruskin next returns to the question of the helm cloud, a topic he first discussed in Volume I. He attempts to modify his earlier theory of their formation which was borrowed from De Saussure, having realised that helm clouds form on what he considers to be hot summits as well as cold. In fact, the formation of helm clouds is related to the wind field and the topography, and not to the temperature of the mountain's surface.[46]

None of Ruskin's later writings after *Modern Painters* contains the same thirst for meteorological knowledge for its own sake. It becomes increasingly difficult to disentangle the symbolism of clouds from the atmospheric science. Hence an examination of the meteorology of *The Storm Cloud of the Nineteenth Century*, written in 1884, will be held over until later in the chapter when we have examined the importance to Ruskin of cloud beauty and the moral significance of clouds.

Overall, it is easy to find fault with the scientific truth of Ruskin's atmospheric writings. Nevertheless, his visual rather than experimental approach to the atmosphere leads a valuable and fascinating trail through the history of meteorology in the second half of the nineteenth century. It is literature that still raises many thought-provoking and as yet inadequately answered questions.

Cloud Beauty: The Imaginative Faculty

The eye provides the truth of landscape form and organisation, but for
Ruskin the argument of the eye must be set against that of the mind —
it is 'mental vision' which penetrates the beauty of landscape. Beauty is
inherent in the landscape itself, in those forms and relations which must
so accurately be observed by the eye, and because of that inherence art
is 'in service' to landscape in its pursuit of beauty. All beauty

> is either the record of conscience, written in things external, or it is a
> symbolising of Divine attributes in matter, or it is the felicity of
> living things, or its perfect fulfilment of their duties and functions.
> In all cases it is something Divine, either the approving voice of God,
> the glorious symbol of Him, the evidence of His kind presence, or
> the obedience to His will by Him induced and supported.[47]

Since beauty lies not within the eye of the beholder, nor arises from
association with human emotions, as had been held by the aesthetic
philosophers of landscape in the eighteenth century, but is to be read in
the landscape itself and in its forms, it is to be found also in the clouds.
Clouds have their particular beauty to which Ruskin's discussion in
Volume V of *Modern Painters* is supposedly addressed, although he
digresses frequently into questions of a scientific and observational
nature. The key to cloud beauty lies in their interplay with heaven's
infinity. The 'cloud-balancings' exist 'to appease the unendurable glory
to the level of human feebleness, and sign the changeless motion of the
heavens with a semblance of human vicissitude'.[48] This theory of cloud
beauty, a perfect example of Ruskin's exegetical approach to landscape,
is elaborated in an extraordinary and at first sight eclectic chapter on
rain-clouds, entitled 'The Angel of the Sea'.[49] Here, Ruskin combines a
theory of environmental determinism with an analysis of Greek
mythology and an interpretation of the nineteenth Psalm.

Initially he divides the world into five regions based upon climate
and vegetation and their fitness for human art.[50] The division is a
simple south to north classification of climatic zones based on
temperature. In the tropics are the 'forest lands' which are 'characterised
by moist and unhealthy heat, and watered by enormous rivers'. They
are unsuited to the cultivation of the mind or to art. The 'sand lands' of
the mid-latitude deserts are the hard environment of the great nomadic
peoples, capable of 'solemn monumental or religious art' but not of a
pleasurable materialist art. The 'grape and wheat lands' of the

Mediterranean form 'the noblest and best ground given to man', and here flourishes a perfect art, combining religious impulse with sensuous materialist pleasure, while in the 'field or meadow lands' of the European plains art loses its spirituality and becomes wholly material. Finally, the 'moss lands' of the north, like the tropical forest lands, he claims can produce no art! The scheme is crude and patently absurd in its posited relationship between climate, environment and art, and it is but one of various such world regional classifications suggested by Ruskin but never sustained beyond their immediate value to him, nor subjected to critical examination.[51] Their purpose is to provide context for the one subdivision thus yielded for his more detailed consideration. In this case it is the 'moss lands' which, he claims, have one great advantage: the sight of the sky which their flatness and lack of vegetation affords, and their direct dependence upon it for their characteristic moss and streams. Physically arduous, offering little comfort to man, these northern lands nevertheless enjoy the constant interplay of sunshine and rain under the influence of what we now recognise as frontal weather.[52] In describing this interplay Ruskin calls forth one of his most powerful landscape descriptions:

> Rain in the moss lands is the messenger sent to a special place on a special errand. Not the diffused perpetual presence of the burden of mist, but the going and returning of intermittent cloud. All turns upon that intermittence. Soft moss on stone and rock; – cave-fern of tangled glen; – wayside well – perennial, patient, silent, clear; stealing through its square font of rough-hewn stone; ever thus deep -- no more – which the winter wreck sullies not, the summer thirst wastes not, incapable of stain as of decline – where the fallen leaf floats undecayed, and the insect darts undefiling. Cressed brook and over-eddying river, lifted even in flood scarcely over its stepping-stones, – but through all sweet summer keeping tremulous music with harp-strings of dark water among the silver fingering of the pebbles. Far away in the south the strong river Gods have all hasted, and gone down to the sea. Wasted and burning, white furnaces of blasting sand, their broad beds lie ghastly and bare, but here the soft wings of the Sea Angel droop still with dew, and the shadows of their plumes falter on the hills: strange laughings and glitterings of silver streamlets, born suddenly, and twined about the mossy heights in trickling tinsel, answering to them as they wave.[53]

The soft rain of the westerlies thus feeds the streams and purifies men. The light which touches the rain-clouds and the blue which Ruskin claims is so frequently visible in patches between them, and is intensified by them, are intended to remind us of God's ever-feeding mercy, both in this life and later in heaven.

The rain-cloud has, however, many faces. The storm cloud for example has a beauty which derives from a different message to men. It manifests divine power and strength. The various forms of rain-cloud Ruskin relates to a family of classical Greek gods, the progeny of a union between Earth and Nereus, the sea-god. Of this progeny the Graii represent the soft rain-cloud already described, while the broad, rushing storm cloud (or the advancing squall front perhaps) is Typhon – the hundred-headed monster with wildly drifting hair so well represented in Turner's *Slavers*. Ruskin notes that this painting, Turner's greatest, was originally described by Turner as 'Slaver throwing overboard the dead and the dying. Typhoon coming on.' Finally, there is the Medusa cloud, the true hail cloud – a turning, twisting, rising vortex which we now know can soar to over 35,000 feet. The three-dimensional air mechanics of such clouds are only today being studied by meteorologists, but Ruskin's close observation of form had revealed this characteristic, and it brought him again to face the vortex form which Turner explored in so many of his later works, including the painting of *Quilleboeuf* at the mouth of the Seine in the *Rivers of France* series, and perhaps most clearly in his *Study of a Ship in a Violent Storm*.

Ruskin had already frequently noted the recurrence of the curving line in nature as evidence for God's mastery in selecting a perfection of form for function.[54] He does not make the same claim for the Medusa vortex and avoids the problem posed by Turner's vortex for his own conventional sense of pictorial representation of truth in landscape. The Medusa hail cloud may indeed rise in a spiral of air movement – it is seen most clearly in the seaspout or tornado, but for Turner the vortex represented a way of breaking through the very perspective conventions which Ruskin was seeking to maintain, allowing Turner to grasp a deeper truth of elemental natural forces in a novel way, and even requiring him to alter the frame shape of his pictures to achieve it.[55] For Ruskin, the curving vortex of the Medusa cloud mediates between the eternal judgement of God in his heaven and his mercy, as represented by the rain-cloud. Therein lies its beauty.

This leads to the final synthesis of cloud beauty, demonstrated in an interpretation of the first lines of the nineteenth Psalm:

> The heavens declare the glory of God
> and the firmament his handiwork.

Ruskin claims that if we approach these lines in a simple way — as was intended by both writer and translator, the 'heavens' quite obviously mean the infinite void of unchanging space beyond the upper reaches of the atmosphere. It is this which signals the inconceivable eternity, glory and unalterable *law* of God. The 'firmament' on the other hand is the clouds which cover the 'unendurable glory' and, in their constant change, manifest the continual presence and ever-flowing mercy through which God makes himself available to men. It is thus that the Medusa cloud, reaching into the heavens from the level of the firmament, judges us and tests our faith, uniting law and mercy.

The beauty of clouds then, in Ruskin's theory, is discovered by close attention to their forms and colours, with the mind applying itself to the information given by the eye. Such beauty has an objective basis in the revelation of the divine. Ruskin's is a deeply *religious* understanding of landscape and its meaning — one wherein the divine is 'neither a justifying explanation of the natural course of the world, nor an abolition and interruption of it: it is itself the natural course of the world.'[56] The task of landscape study and particularly landscape painting is thus 'to declare the perfectness and eternal beauty of the work of God; and [to] test all work of man by concurrence with, or subjection to, that'.[57]

Moral Significance

Although the moral purpose of art was never questioned by Ruskin, his teleological understanding of nature changed through time. His evangelical upbringing led to the purpose of the early volumes of *Modern Painters* being to 'understand that God paints the clouds . . . that men may be happy in seeing Him at His work'.[58] However, even as early as 1851 he wrote to Henry Acland that his evangelical faith 'flutters in weak rags . . .' By 1860 Ruskin had experienced a 'religious unconversion' that changed his central concern from God-in-nature to God-in-man. From 1860 onwards Ruskin turned his attention to society. His attacks on conventional economics, however, as Hewison states, 'did not resolve the problem of the contradictory need for good art to make society noble, and for a good society to produce great art'.[59] Ruskin became disillusioned and turned to other outlets such as *The Queen of the Air* as a form of escapism. First published in 1869, and subtitled 'A Study of the Greek Myths of Cloud and Storm', the work is the story of Athena:

This great goddess, the Neith of the Egyptians . . . the Minerva of the

Latins, is, physically, the queen of the air; raising supreme power both over its blessings of calm, and wrath of storm; and spiritually, she is the queen of the breath of man . . .[60]

Athena was for Ruskin 'the spirit of life',[61] although it was around this time that his own spirit for life began to flag. In the Preface to *The Queen of the Air* he writes:

I have seen strange evil brought upon every scene that I best loved, or tried to make beloved by others. The light which once flushed those pale summits with its rose at dawn, and purple at sunset, is now umbered and faint; the air which once inlaid the clefts of all their golden crags with azure is now defiled, with languid coils of smoke, belched from worse than volcanic fires . . .[62]

Suddenly the atmosphere was blotted out and defiled by a new type of cloud – industrial and domestic smoke. Levels of smoke and sulphur dioxide peaked in London and the rest of Britain around 1880.[63] Much has been made of the inverse relationship between the increasing pollution and Ruskin's declining psychological state. Hewison writes:

The serpentine coils of smoke blew up into a black cloud, which . . . began to blot out Ruskin's perception of God in Nature. At first the cloud is associated with disappointment and failure . . . [and then] this new form of Mountain Gloom is our active force of evil.[64]

But the smoke was real enough, as Ruskin angrily notes in his Brentwood diary for 13 August 1879:

The most terrific and horrible thunderstorm, this morning, I ever remember . . . the air one loathsome mass of sultry and foul fog, like smoke . . . It lasted an hour, then passed off . . . Settling down again into Manchester devil's darkness.[65]

The meteorology of *The Storm Cloud of the Nineteenth Century* is more advanced than in any of his previous writings. For instance, he now distinguished two types of rain-cloud – the electric and the non-electric. However, because it was originally presented as a lecture, Ruskin deliberately simplified his explanations. There is not space here to match Ruskin's diary observations of 'The Plague-Wind' and 'The Storm-Cloud' with actual meteorological observations; suffice it to say that current meteorological theory has confirmed strong links between pollution and weather. Smoke and other pollutants markedly affect urban climates.[66]

Ruskin's claim that he had discovered a new type of cloud was of course a myth — Athena raped by political economy. The storm cloud is the ultimate stamping of Ruskin's personality on meteorology. As Kirchoff writes:

> The glamor of conventional science lies in its intellectual dominance of Nature. Sensing the alienation implicit in this dominance, Ruskin denies himself Francis Bacon's appeal to the urge for power. He must present a way of knowing Nature that implies its opposite: the final elusiveness of all natural phenomena . . . To question the primacy of abstract 'Laws', he substitutes the cantankerous personality of John Ruskin.[67]

Ruskin was not the only one to be disgusted by man's defilement of nature. As Rosenburg states:

> 'Modern Painters' and 'The Storm-Cloud' recapitulate, at both ends of his career, one of the crucial realizations of his century: that the benign Nature of Wordsworth was in fact red in tooth and claw. With that discovery, the pastoral landscapes of the Romantic poets gave way to the sinister, blighted terrain of Tennyson's 'The Holy Grail', of Browning's 'Childe Roland', and of their prose counterparts, Hardy's blasted heaths and Ruskin's 'Storm-Cloud'.[68]

Conclusion

The storm cloud of the nineteenth century may not have been a meteorological reality, but the moral imperative which made it real for John Ruskin points to issues which transcend empirical veracity. The collection of which this essay is a part testifies to geography's recognition that the imaginative treatment of landscape and place provides a rich vein of meanings often unavailable elsewhere. It is important therefore to place Ruskin's cloud writings not merely in the context of prevailing knowledge about meteorological phenomena and processes, but in the context of nineteenth-century science generally and of related attitudes to man and society. Insofar as Ruskin's protests and vision were directed against a view of nature and society which still

holds considerable currency in the modern world, his critique may have
a contemporary relevance.

We have shown that Ruskin's meteorological knowledge was suspect,
that he held a conservative view of pictorial representation and that his
understanding of natural beauty was grounded in a teleology from which
he derived a didactic purpose for art. The ideas that nature is the manifest handiwork of God, and that art has a moral purpose rather than a
purely aesthetic one, are by no means confined to John Ruskin. Yi-Fu
Tuan, for example, has demonstrated the profound influence of
teleology upon the intellectual struggle towards understanding the
hydrological cycle in the early modern age.[69] Yet Ruskin does not
regard nature as a machine in the manner of what Collingwood has
termed the 'Renaissance Cosmology'.[70] He is enough of a nineteenth-
century man, enough aware of the development of science, to recognise
that there are laws of development inherent in nature itself; and it is to
nature that he looks for much of their explanation. But his is a
curiously uneasy position. He is unable to accept fully the modern idea
that sufficient explanation of these laws lies within nature, and that
many of them are invisible to the eye and unamenable to biblical
interpretation. His own penetrating observation constantly forces him
into questions which he must declare as beyond the intended scope of
human knowledge, only to be shown later that they can be answered by
material science. His problem with the floating of clouds is but one
example. The spectacular progress of nineteenth-century science in
explaining natural processes in a materialist way consistently
undermined Ruskin's central thesis, but he refused to reject the central
moral certainty that a *human* world could not be constructed from a
purely materialist premiss. His rejection of Darwin's shattering
demonstration of evolution in organic life is based upon this moral
certainty. It is a revulsion from the human implications of Darwinism
which he shared with many religious minds of his age, and it surely
accounts in part for his final despair in landscape as a unity of earth, air
and human endeavour:

> That harmony is now broken, and broken the world around:
> fragments still exist, and hours of what is past still return, but month
> by month the darkness gains upon the day, and the ashes of the
> Antipodes glare through the night.[71]

Nor is the notion of a didactic purpose of art peculiar to Ruskin.
Nikolaus Pevsner has argued that it is characteristic of English art over a

long period.[72] But Ruskin's assertion of a moral purpose for art is not
located at the level of general spiritual uplift, it is directed specifically
at his contemporaries and is used as a vehicle for a frontal attack upon
the assumptions of nineteenth-century political economy. He counter-
poses the function of art in society to a view of society's motivation
derived from the classical economic theory of Adam Smith and John
Stuart Mill. Their view of a society based upon the aggregate of
individuals' pursuit of maximum utility, he claims, reduces man to a
mere brute and destroys any moral foundation to society. No true art
could flourish in such conditions. In this, as the Marxist art historian
Arthur Hauser has pointed out,

> he was indubitably the first to interpret the decline of art and taste
> as the sign of a general cultural crisis, and to express the basic, and
> even today not sufficiently appreciated, principle that conditions
> under which men live must first be changed, if their sense of beauty
> in the contemplation of art is to be awakened.[73]

But here again Ruskin's position is ambiguous. His criticism of
nineteenth-century political economy, essentially of capitalism and its
consequences, remains a *moral* one. He does not accept the Marxist
examination of the internal dynamics of capitalist social development
any more than the assumptions it attacked, for to him both are
grounded in the same materialism.

Similarly, his recognition of the need to express through landscape
art a more profound truth than is possible in the mere accurate
rendering on canvas of the forms of landscape, is always arrested short
of breaking through the pictorial conventions inherited from the
Renaissance, and leaves him confused in the face of Turner's most
powerful landscapes and dismissive of the revolutionary depiction of
atmosphere pioneered by the Impressionists.

There is, however, a sense in which it is precisely Ruskin's ambiguity
and uncertainty in the face of nineteenth-century thought and
nineteenth-century reality, his hovering between a materialist and an
idealist religious position, that renders his writing relevant today. The
alienation of humans from each other and from the natural world,
against which Ruskin protested in the 1880s, is seen by many to have
progressed much further by the 1980s. Over the past decade, however,
the optimistic and prescriptive materialism which has characterised
both popular and academic thought for a century has been radically
challenged. A great deal of popular protest echoes Ruskin's appeal for a

reformulated harmony of man and nature. It does so on a basis of ideas which are as apparently incompatible as those with which Ruskin grappled. On the one hand it appeals to 'scientific' laws of ecological interdependence (the outcome of Darwinian theory), and on the other to anti-scientific mysticism, often of an Eastern religious variety. These are unified by moral conviction rather than logical relations. The search for meaning in place, and a protest against environmental defilement, represent something of humanistic geography's contribution to that challenge.

Similarly, a revived Marxism is directing much of its intellectual energy towards incorporating a sense of the significance of human sensuousness and creativity into what too frequently has become a reductionist thesis, wherein laws of material economic development in the 'base' of society have been read into the cultural 'superstructure' in a crude fashion, unmediated by moral considerations. Ruskin offers much to the meteorologist and geographer of the 1980s. His recognition that the scientific and imaginative study of clouds and landscape cannot be divorced from the issues of man's humanity, offers at the very least a set of questions about the relations of science and art in society, and the meaning of landscape for men — and some of the terms within which the discourse of geography may proceed.

Notes

1. A perfect example from literature is the constant counterposing of equable and stormy weather conditions as metaphors for the contrasted characters of the main protagonists of Emily Brontë's *Wuthering Heights.*
2. D. Cosgrove, 'John Ruskin and the Geographical Imagination', *Geographical Review*, vol. 69, no. 1 (1979), pp. 43–62.
3. P. M. Walton, *The Drawings of John Ruskin* (Clarendon Press, Oxford, 1972), pp. 16–17.
4. J. E. Thornes, 'Constable's Clouds', *The Burlington Magazine*, vol. 121, no. 920 (1979), pp. 679–704; R. Rees, 'John Constable and the Art of Geography', *Geographical Review*, vol. 66, no. 1 (1976), pp. 59–72.
5. E. Soreni, *Storia del paesaggio agrario Italiano* (Latuza, Bari, 1974).
6. See Ruskin's attack on such 'ideal' landscape in the preface to the second edition of *Modern Painters I* (*Works*, vol. 3, pp. 7–52). Volumes of the complete works of John Ruskin will be referred to in this way in the remaining notes. They were edited by E. T. Cook and A. Wedderburn (39 volumes) and published by George Allen, London, and Longmans, Green and Co., New York, 1903–12.
7. W. A. M. Peters, *Gerard Manley Hopkins: a critical essay towards the understanding of his poetry* (Oxford University Press, London, 1948), p. 2.
8. For example R. Hartshorne, *The Nature of Geography* (Association of American Geographers, Lancaster, Pa. 5, 1961), pp. 149–74.
9. C. O. Sauer, 'The Morphology of Landscape' in J. Leighly (ed.), *Land and*

Life (University of California Press, Berkeley and Los Angeles, 1963), p. 345.

10. J. Leighly, 'Some Comments on Contemporary Geographic Method', *Annals, Association of American Geographers*, vol. 27, no. 3 (1937), pp. 125–41.

11. A. Buttimer, *Values in Geography* (Association of American Geographers, Resource Paper No. 24, 1974).

12. E. Panofsky, *Meaning in the Visual Arts* (Penguin, Harmondsworth, 1970), p. 65.

13. J. Ruskin, *Lectures on Art*, 1870 (*Works*, vol. 20, p. 107).

14. J. Ruskin, *Lectures on Landscape*, 1871 (*Works*, vol. 22, p. 12).

15. J. Ruskin, *Praeterita*, 1886 (*Works*, vol. 35, p. 54).

16. J. Evans and J. H. Whitehouse, *The Diaries of John Ruskin* (Oxford University Press, London, 1956), p. 2.

17. J. Ruskin, 'Remarks on the Present State of Meteorological Science', *Transcriptions of the Meteorological Society*, vol. 1 (1839), pp. 56–9. (*Works*, vol. 1, pp. 206–10.)

18. Ruskin, *Modern Painters I* (*Works*, vol. 3, p. 352). As we shall see, the chapters on cloud beauty in *Modern Painters V* contain more physics in fact than the chapters on cloud truth in *Modern Painters I*.

19. It is not known whether Ruskin had read any books on meteorology before he published the first volume of *Modern Painters* in 1843. Ruskin quotes De Saussure in later volumes and may well have read De Saussure's *Essais sur l'hygrometrie* (1783) which was based chiefly on observations of mountain weather.

20. Ruskin, *Modern Painters I* (*Works*, vol. 3, p. 346). This explanation is due to De Saussure.

21. Lord Rayleigh, 'On the Light from the Sky, its Polarisation and Colour', *Philosophical Magazine*, vol. 41 (1871), pp. 107–20 and 274–9.

22. J. Ruskin, *The Queen of the Air*, 1869 (*Works*, vol. 19, p. 292).

23. Ruskin, *Modern Painters I* (*Works*, vol. 3, p. 352).

24. In what is considered by many to be the foremost book on the subject – M. Minnaert, *Light and Color in the Open Air* (Dover Editions, New York, 1954) – Ruskin is quoted ten times.

25. L. Howard, *On the Modification of Clouds* (J. Taylor, London, 1803).

26. Ruskin, *Modern Painters I* (*Works*, vol. 3, p. 359).

27. Howard, *On the Modification of Clouds*, p. 8.

28. See F. H. Ludlam, 'History of Cloud Classifications' in R. S. Scorer (ed.), *Clouds of the World* (David & Charles, Newton Abbot, 1972), p. 17.

29. Goethe was enchanted by Howard's cloud classification, it provided the 'missing thread' in his conception of the relationship between man and God. As K. Badt notes in *John Constable's Clouds* (Routledge & Kegan Paul, London, 1950), p. 19, for Goethe:

> The lower sphere of the stratus, of the streaks of mist, is nature and only nature; the middle region of the cumulus clouds on the other hand is already understood in the sense of a higher atmosphere, to attain to which it is necessary to possess a special ability. In the highest layer of air, however, that in which the cirrus cloud hovers, we see, as it were, the veiled image of the intellectual and the spiritual.

Goethe was moved to write a poem, 'In Honour of Howard', which has verses devoted to each of the main cloud types of Howard, a translation of which is given in D. Scott, *Luke Howard* (William Sessions, York, 1976), p. 26, from which the verse on cumulus reads:

Still soaring, as if some celestial call
Impell'd it to yon heaven's sublimest hall;
High as the clouds, in pomp and power arrayed,
Enshrined in strength, in majesty displayed;
All the soul's secret thoughts it seems to move,
Beneath it trembles, while it frowns above.

30. Ruskin, *Modern Painters I* (*Works*, vol. 3, p. 361). Altostratus and altocumulus are cloud names first suggested by Renou in 1855.

31. Ibid., p. 370.

32. Ibid.

33. Ibid., p. 376.

34. G. Hough, *The Last Romantics* (Duckworth, London, 1949), p. 9.

35. E. Lehrs, *Man or Matter* (Faber & Faber, London, 1951), p. 119.

36. As noted in C. R. Leslie, *Memoirs of the Life of John Constable*, 2nd edn (Longman, London, 1845), p. 350.

37. Ruskin, *Modern Painters I* (*Works*, vol. 3, p. 399). In a footnote mamma clouds are accurately described: 'It is seen chiefly in clouds gathering for rain, when the sky is entirely covered with a grey veil rippled or waved with pendant swells of soft texture, but excessively hard and liny in their edges.'

38. J. Ruskin, *Eagle's Nest*, 1872 (*Works*, vol. 22, p. 211).

39. J. Ruskin, *Modern Painters V*, 1860 (*Works*, vol. 7, p. 135). Ruskin probably knew of the medieval poem *The Cloud of Unknowing* (Penguin edition introduced by C. Wolters, first published 1961) which beautifully sets out to show that man's intellect is not capable of understanding the works of God.

40. The theory was finally demolished by Kober, see W. E. Knowles Middleton, *A History of the Theories of Rain* (Oldbourne, London, 1965), p. 56.

41. Ruskin, *Modern Painters V* (*Works*, vol. 7, p. 137).

42. Quoted in the preface of *Works*, vol. 7, p. lxi.

43. J. Ruskin, *Coeli Enarrant* (George Allen, London, 1885).

44. Ruskin, *Modern Painters V* (*Works*, vol. 7, p. 152).

45. H. Damisch, *Théorie du Nuage Pour une Histoire de la Peinture* (Editions du Seuil, Paris, 1972).

46. Scorer, *Clouds*, p. 71.

47. Ruskin, *Praeterita* (*Works*, vol. 35, p. 224).

48. Ruskin, *Modern Painters V* (*Works*, vol. 7, p. 133).

49. Ibid., p. 175.

50. Ibid., p. 177.

51. Ruskin uses another classification for example in *The Poetry of Architecture*, written in 1837 and discussed in Cosgrove, 'John Ruskin', pp. 47–8.

52. There appears to be a slippage here in Ruskin's understanding of the geographical location of the 'moss lands'. Whilst he mentions Sweden as his example, what he appears to have had in mind during the discussion of their climate is the western and northern uplands of Britain, characterised in their weather by the passage of frontal systems.

53. Ruskin, *Modern Painters V* (*Works*, vol. 7, p. 178).

54. Cosgrove, 'John Ruskin', pp. 55–6.

55. J. Lindsay, *Turner, His Life and Work* (Panther, St Albans, 1966), p. 270.

56. W. J. Otto, *The Homeric Gods* (Pantheon Books, New York, 1954), quoted in V. Scully, *The Earth, The Temple and the Gods: Greek Sacred Architecture* (Yale University Press, New Haven and London, 1962), p. 7.

57. Ruskin, *Modern Painters V* (*Works*, vol. 7, p. 9).

58. J. Ruskin, *Modern Painters III*, 1856 (*Works*, vol. 5, p. 384).

59. R. Hewison, *John Ruskin: The Argument of the Eye* (Thames & Hudson, London, 1976), p. 148.

60. Ruskin, *Queen of the Air* (*Works*, vol. 19, p. 305).

61. Ibid., p. 346.

62. Ibid., p. 293.

63. P. Brimblecombe, 'London Air Pollution 1500–1900', *Atmospheric Environment*, vol. 11 (1977), pp. 1,157–62.

64. Hewison, *John Ruskin*, p. 159.

65. J. Ruskin, *The Storm Cloud of the Nineteenth Century*, 1884 (*Works*, vol. 34, p. 37).

66. J. E. Thornes, 'London's Changing Meteorology' in H. Clout (ed.), *Changing London* (University Tutorial Press, London, 1978), pp. 128–37.

67. F. Kirchoff, 'A science against sciences: Ruskin's Floral Mythology' in U. C. Knoepflmacher and G. B. Tennyson (University of California Press, London, 1977), p. 247.

68. J. D. Rosenburg, *The Genius of John Ruskin* (Allen & Unwin, London, 1963), p. 322.

69. Yi-Fu Tuan, *The Hydrological Cycle and the Wisdom of God* (University of Toronto Press, Toronto, 1968).

70. R. G. Collingwood, *The Idea of Nature* (Clarendon Press, Oxford, 1945).

71. Ruskin, *Storm Cloud* (*Works*, vol. 34, p. 78).

72. N. Pevsner, *The Englishness of English Art* (Penguin, Harmondsworth, 1964).

73. A. Hauser, *A Social History of Art* (2 vols., Routledge & Kegan Paul, London, 1952), quoted in G. P. Landow, *The Aesthetic and Critical Theories of John Ruskin* (Princeton University Press, Princeton NJ, 1971), p. 10.

3 LITERATURE AND 'REALITY': THE TRANSFORMATION OF THE JUTLAND HEATH

Kenneth Robert Olwig

There has been a tendency in recent geographical writing on literature to focus upon the presumed perspicuity of the individual author. Geographers write of the 'heightened perception of the artist and his ability to communicate that experience'.[1] This chapter, however, is concerned not so much with the individual's apprehension of geographic reality as it actually *is*, but with literature's social function in envisioning reality as it *is not* but *ought to be*, and with its potential, thereby, for stimulating change. This shift in emphasis is related to a theoretical stance which questions the focus upon the perceptual abilities of the individual. The significance of the distinction between these two approaches to literature will be illustrated by the case of the nineteenth-century transformation to forest and field of the heaths of Jutland, Denmark, as seen in the light of several recent geographical literary studies. A theoretical introduction will therefore be followed by a discussion of the aspects of these recent studies which are pertinent to the perspective on literature and geography which is suggested by the case of the Jutland heath.

Literature and Geography

C. L. Salter and W. J. Lloyd have outlined two, 'not mutually exclusive', geographical approaches to landscape in literature.[2] They are, in fact, complementary in that the first focuses on literature's presumed ability to reproduce the objective qualities of landscape whereas the second emphasises the subjective experience of that landscape. 'On the one hand', they write, 'authors may be respected for their accurate reproduction of an objective landscape', for their abilities as 'keen observers'. On the other hand, authors are singled out for their 'sensitive insights into the subjective' and their ability to 'articulate human experience'.[3] A classic example of the view that literature can reproduce objective reality is that held by E. W. Gilbert (who sees literature as the last bastion of descriptive regional geography):

many novels present life and work on a clearly marked piece of land
with truth. Reality is faithfully shown: it is not lost in the dim
twilight of modern geographical jargon. The novelists paint a picture
of real earth . . .[4]

The complementary view which stresses the subjective human experience
of that same objective reality is expressed in the statement by David
Seamon:

> By uncovering and understanding environmental experience as other,
> more sensitive and creative individuals have known it, we, as more
> typical people, may become aware of patterns in our *own*
> experience that we had not known before. In this way we
> ourselves become more sensitive to our *own* geographical situation.[5]

These two approaches are complementary, and not mutually
exclusive, because they deal with the objective and subjective perception of the same reality, *as it is*. There is, however, a fairly widespread
view of literature, and art in general, which sees the relationship of
literature to reality in a very different way. This view emphasises not
the perceptive sensibilities of the individual artist, but the autonomous
role of art itself (*artistic form*) as a 'social fact and as a mode of
communication', to use the words of Northrop Frye.[6] Art, from this
perspective, furthermore, is not fundamentally concerned with the
presentation of a heightened, more vivid perception or experience of
reality as it actually is. Art, quite the opposite, due to its very form, is
seen as being basically 'estranged' from this reality. To illustrate this
view the arguments of a number of philosophers and critics may be
drawn upon; notably the philosophers Susanne Langer and Herbert
Marcuse, and the critic Northrop Frye.

   For Langer the central interest of art lies in its 'uncoupled estrangement' from 'actuality' through being given artistic form, which creates a
'self-sufficient' realm of 'illusion'.[7] Marcuse takes a similar position
when he argues that it is through the 'transformation of a given content
(actual or historical, personal or social fact)' into the 'self-contained
whole' which he calls 'aesthetic form', that the work is ' "taken out" of
the constant process of reality and assumes a significance and truth of
its own'.[8] The consequence of this view is that art is credited not with
presenting a different *perception* of reality, but an altogether *different*
reality:

What is at stake is the vision, the experience of a reality that is so fundamentally different, so antagonistic to the prevailing reality that any communication through the established means seems to reduce this difference, to vitiate this experience.[9]

This view, of course, does not deny that artists can seek to create the illusion of factual reproduction, as in *trompe l'oeil* painting or modern 'New Wave' French poetry.[10] This, however, is not the basic function of art, which, by transforming a subject into the form of a painting or poem (in this case in the style of a super-realistic convention), makes its subject into something quite 'other'. It is this transformation, which occurs in all art, which is of primary interest, and not the occasional apparent attempt to create the illusion of fact. Though the illusion may be enormously realistic, this only increases the credibility of the fundamentally different reality of art, and the social scientist, as will be seen, must be particularly wary of its deceptive power.

Though art may be 'uncoupled' from actuality, this does not exclude it from having an impact upon society's perception and experience of that actuality. Quite the reverse. Its impact lies in this 'uncoupling', which, in Langer's words 'liberates perception', thereby enabling its public to experience actuality from a fresh perspective.[11] Marcuse likewise argues that it is artistic form which 'gives the familiar content and the familiar experience the power of estrangement – and which leads to the emergence of a new consciousness and a new perception'. Art is thus able to 'break the monopoly of established reality'.[12]

Art can establish this 'break' by creating the image of a fundamentally different reality which, in fact, as Marcuse writes, 'contains more truth than does everyday reality'.[13] It is reality not as it *is*, but as it *ought to be*. Frye makes a similar argument when he characterises 'the archetypal function of literature' as being the visualisation of the 'world of desire, not as an escape from "reality", but as the genuine form of the world that human life tries to imitate'.[14] This 'archetypal function' is well illustrated by a form of art which is generally acknowledged to have a significant impact upon our perception of both rural and urban landscapes, that of the so-called 'pastoral convention'.

It is particularly through the work of the literary scholars Leo Marx and Raymond Williams that the importance of the pastoral convention has come to the attention of geographers.[15] (Strictly speaking, the term 'pastoral' should only be applied to art which is primarily concerned with the life or environment of pastoralists, but it is also common to apply it, as both Marx and Williams do, to related art which deals more generally with rural life. The term will thus also be used in this broader sense here.) In this pastoral convention, which can be traced back to classical times, an ideal rural world of peace — associated with innocence of the childhood past of both society and the individual — is counterposed to an urban existence which is its obverse. The essence of this opposition, and its impact upon perception, is captured in this statement by Williams:

> Country life has been an image and even a common-place of natural virtue and order in many literatures . . . Often it's the simple image: of the peace and quiet of the country as against the ugly mechanism of the industrial town. But country life has also been seen repeatedly as the life of the past: of the writer's childhood, or of man's childhood, in Eden and the Golden Age. And it is here that the image becomes confused with history, so that many writers come to set dates, periods and historical formulations to the long habit of rural retrospect.[16]

Marx and Williams both show how the perception of the country and the city by generations of social thinkers has been influenced by the pastoral. 'There has been', Williams writes, 'a failure to read literature as literature, and then the making of a false history of a false reading.'[17] The problem is that they have been deceived by the illusion, created by the author through the medium of this form of art, that the 'world of desire' of the rural idyll is not 'an escape from "reality"', but the genuine form of the world'. Geographers clearly have reason to be wary lest neophyte assumptions about the objectivity of literature lead them to 'read literature as geography', and thereby 'make a false geography of a false reading'. This is no reason, however, to abandon the geographical study of literature. On the contrary, an understanding of the process by which such false geographies are created is of considerable importance to an understanding of geographical perception and experience. The value of learning to 'read literature as literature' can be shown by reference to several examples.

Literature and Landscape 'Reality'

A Romantic poet such as Wordsworth sold few copies of his works in the Lake District. Romantic poets tended to write about the landscapes of the Lake District, the Scottish Borders, Wales or (to depart Britain) Switzerland, which were at the social and economic periphery of an urbanising and industrialising Europe. The public for this poetry, however, was at the 'core', where the need to 'uncouple' from the actuality of its often grim environment was greatest, and it is against this background that the popularity of this poetry must be seen. The natural environments portrayed in this poetry helped break the 'monopoly' of the established reality of towns by presenting a reverse image of their unnatural qualities, rather than a faithful depiction of the reality of these rural districts. Jane Zaring has documented in her study 'The Romantic Face of Wales' how artists systematically neglected to describe evidence of industrialisation and urbanisation in the landscapes which they seem to depict so faithfully.[18] This myopia was in turn reflected in the experience of tourists whose perception of the area was influenced by Romantic art.

If Romantic poetry did not faithfully represent the actuality of such landscapes, it may have accurately portrayed their ideal qualities so that they appeared as a more 'genuine form of the world'. Romantic poets often made clear that they saw their task as being the preservation of the memory of the last ideal vestiges of an earlier society now threatened by the expansion of urban commerce and industry. They were thus concerned not so much with the reality of these areas as it actually was, but with the remnants of their character as they supposed it to have been and as they believed it ought to be. Though a writer such as Wordsworth made a serious effort to give historical geographic evidence for the 'perfect republic of shepherds and agriculturalists' which he saw as having existed in the Lake District, this nevertheless fits a pattern of perception which was a well-established part of the pastoral convention.[19] It is an approach to such fringe areas which can be traced at least as far back as Virgil's idealisation of pastoral Arcadia, a hilly peripheral district of Greece which he described as the place where the last vestiges of an earlier golden age rural paradise were to be found. Virgil likewise found in these vestiges of the past an image of what reality ought to be, which he then projected into the future as a coming golden age. These ideal qualities of the past thus become the vehicle for a new consciousness of how things could be.

Just as the pastoral convention tends to exaggerate the idyllic qualities of country life, thus providing a foil against which to counterpose the unnatural life of the town, it likewise tends to heighten the negative qualities of the town. If the rural ideal is to appear not as an 'escape', but as 'the genuine form of the world', then the actual urbanised world must be 'inverted', as Marcuse puts it, so that 'it is the given reality, the ordinary world which now appears as untrue, as false, as deceptive reality'.[20] It is not the ideal realm of art which should appear as illusion, but the 'given reality'. 'Illusion' thus comes to be seen, in Frye's words, as 'whatever is fixed or definable, and reality is best understood as its negation: whatever reality is, it's not that'.[21] The actual urbanised world is thereby made to appear as the false reality, the inverse of the true reality of the rural idyll.

The apparent influence of the pastoral convention upon the literary perception of urban industrialised regions is illustrated in D. C. D. Pocock's study of the novelist and the North of England.[22] Through numerous well-chosen examples Pocock documents how from the time of Charles Dickens to the present novelists have consistently portrayed the region as being as devoid, in effect, of rural charm as Wales has been seen to have been its repository. D. H. Lawrence's depiction of the Northern town of 'Tevershall' is typical:

> It was as if dismalness had soaked through everything. The utter negation of natural beauty, the utter negation of the gladness of life, the utter absence of the instinct of shapely beauty which every bird and beast has, the utter death of the human intuitive faculty was appalling . . . ugly, ugly, ugly.[23]

Whatever reality is, it cannot be this! Such a depiction shocks the reader into uncoupling from and reflecting upon that reality which society seems to have created. The artwork thereby creates the basis for a new perception and a new consciousness, which, in this particular case, can be seen to have affected English social policy.[24]

The apparent realism of literary descriptions of urban squalor lends to their effectiveness, but we should not be deceived into assuming that this is any more faithful to fact than was 'The Romantic Face of Wales'. The North, according to Pocock, continues to be depicted in the tradition of Dickens even though this does not reflect the reality of the North. This, of course, is to the dismay of those concerned with boosting the region's image. They no doubt know all too well how literature can affect the patterns of people's experience, regardless of the reality of a place.

It is not just the urban North which literature has perceived as the inverse of the rural idyll; the very countryside itself was its own negation. Witness this passage by Arnold Bennett:

> the vaporous poison of their ovens and chimneys had soiled and

shrivelled the surrounding country till there is no village lane within a league but what offers a gaunt and ludicrous travesty of rural charms.[25]

The characterisation of rural existence as the inverse of what is to be expected given the pastoral convention has been termed 'counter-pastoral' by Williams.[26] The contrast between the rural idyll as it ought to be and the desultory realm of the counter-pastoral creates a sense of outrage which tends to be directed against the alien forces which have been responsible for the desecration of this natural state. It can also, however, be used to create a feeling of elegiac mourning for a lost past.

A particularly well-known British example of the counter-pastoral is Oliver Goldsmith's 'The Deserted Village'. Here, enclosure has created a landscape which is figuratively described as 'a grave', its population sent into exile in the wilderness. Though this poem has often been assumed to be a reasonably faithful depiction of conditions in Goldsmith's home village of Lissoy in the eighteenth century, much of it, like the contrasting scenes of innocent country sports and those of exile with its graveyard symbolism, were standard features of the pastoral convention, and can be found as far back in time as the *Eclogues* and *Georgics* of Virgil. One must not assume, however, that the application of conventional forms to an actual British social issue vitiated Goldsmith's message. Quite the opposite. 'The problem of convention', as Frye puts it, 'is the problem of how art can be communicable.' The convention, in this case, places individual events in a particular British rural district in a larger framework which resonates with social and cultural associations penetrating far back into history. By allowing him to use, in Frye's words, 'a highly concentrated metaphorical language without any breach of decorum', this convention aided Goldsmith in what was apparently his true endeavour.[27] This was the making of a thoroughly researched commentary on the effects of commerce and inequality upon rural Britain.[28]

Literature, then, can be seen to portray landscape reality, not as it is or was, but rather on the basis of a conception of what it ought to be: a conception explicable in terms of the pastoral convention. When literature does achieve a faithful depiction of reality in this wider sense of what it ought to be and can become, then it may indeed have some impact upon what reality becomes, although it cannot, of itself, change reality. The transformation of the Jutland heath is a notable example of a case where literature appears to have been successful in depicting reality in this wider sense, thereby influencing the course of landscape change. It is to this example in greater detail that the chapter will now turn.

The Transformation of the Jutland Heath

The story of the transformation of the Jutland heath has often been told because it is a classic example of regional development. Figures 3.1 and 3.2 depict that dramatic transformation, while the following quotations from Thorpe and from Fullerton and Williams encapsulate the standard view of what occurred. Thorpe wrote:

> By far the greatest concentration of heath was to be found in Jutland, particularly the west and north, where it formed a characteristic landscape covering approximately three million acres in 1800, or about 40 per cent of the entire area of the peninsula . . . With the foundation of the Danish Heath Society in 1866, with its headquarters first at Aarhus and later at Viborg, a really determined attack on the heath began. By 1950 the extent of heath, dune and bog in Jutland had been reduced to 640,000 acres, representing only 8.8 per cent of the peninsular area.[29]

The joint work recorded:

> the demand for new land rose especially after the loss of North Slesvig in 1864. Bravely and industriously, the local population waged war on the heather, encouraged after 1866 by the Danish Heath Society (Det danske Hedeselskab). How successful the reclamation has been is testified by the well laid out farms and plantations which now cover the outwash plains and the adjacent poorer moraines. After a century, over four-fifths of heathland have been converted to agriculture and forestry.[30]

This view is essentially correct, but it leaves a number of questions unanswered. Why, for example, would a people be motivated to 'wage war' and 'make a determined attack' upon a peaceful landscape which many regarded as beautiful? This is a particularly difficult question when it is known that the leading agronomists of the time opposed the cultivation of the heath as a dubious agricultural and economic undertaking.[31] Why, then, did the nation gather behind an effort to

Figure 3.1: The Extent of Jutland Heath about 1800

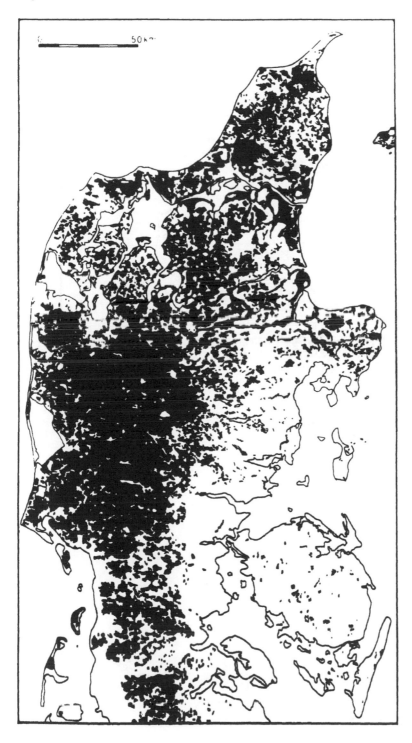

Figure 3.2: The Extent of Jutland Heath about 1950

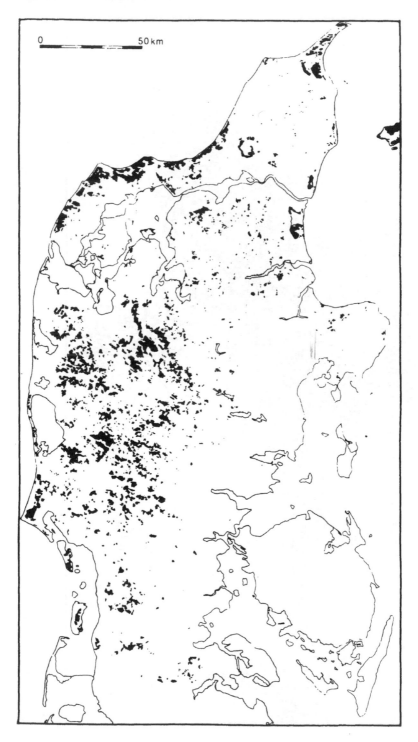

transform the treeless, barren landscape of the heath into a region of forests and fields? An examination of the 'literary landscape' of the heath can help provide the answer.

The Literary Landscape of the Heath

When Hans Christian Andersen wrote his popular poem 'Jutland Between Two Seas' in 1859 there were few in the nation who believed that the heath either could or would be cultivated. Influential scientists, in fact, believed the area to have been the result of natural causes and inherently destined to its bare, forestless condition.[32] Nevertheless, Andersen's poem, written during a visit to Jutland, included the following verses:

> The heath, yes, it is hard to believe —
> but come yourself, look it over:
> the heather is a splendid carpet
> flowers crowd for miles around.
> Hurry, come! in a few years
> the heath a grainfield will be.

> Jutland between two seas
> Lying as a runestone,
> Your future unfolds your power;
> the sea with all its breath
> sings the loudest of Jutland's shores.[33]

To someone who is not a Dane it can be difficult to interpret the meaning of Andersen's linkage of runestones and graves with an image of a future which most thought to be impossible. A clue, however, is to be found in Andersen's diaries where he relates that he was inspired to visit the heaths by the authorship of one Steen Steensen Blicher (1782–1848). The concept of 'Blicher's Jutland heath' is a Danish equivalent to 'Wordsworth's Lake District', 'Scott's Border Country' or 'Hardy's Dorset'. It was Blicher who virtually created the nation's image of the heathland districts of central and western Jutland.[34] When seen in the light of Blicher's authorship, Andersen's symbolism becomes a good deal more intelligible.

'Blicher's heath' was essentially the landscape of the counter-pastoral and its elegiac variant. In his landscape, or rather, heathscape descriptions, Blicher consistently played upon the contrast between the rural idyll and the wild, barren 'desert' of the heath. At times he portrayed

the heath as a 'gaunt', harsh 'travesty' of rural charms, where blowing
sand and sea fog replace the soot and vaporous poison of the literary
landscape of England's North, or, for that matter, the landscape ravaged
by the effects of storm or war which can be found in the poetry of
Goldsmith or Virgil. He was equally at home, however, in the
melancholy mood of the elegy to a lost past, inspired by its graves.
Whereas in the classical elegy these graves are marked by the familiar
marble monuments and tombs, Blicher was able to capitalise on the
manner in which the bare, dark heath tended to make its many ancient
barrows and monoliths stand out in the landscape. The heath thus came
to be characterised as the place of 'our forefathers' dark brown burial
fields', its barrows the 'heather houses of the dead'.[35] By contrasting
the sombre environment of the heath with the conventions of the idyll,
with its associations with childhood innocence, he was able to invest it
with a sublime beauty which both attracted and repelled, while
directing the reader's thoughts to a more ideal past:

> My birthplace is the heather's brown land,
> my childhood's sun has smiled on the dark heath,
> my infant foot has walked the yellow sand,
> among black barrows lives my youth's happiness.[36]

Blicher perceived the heath as literally preserving the remnants of a
more ideal past. The patches of twisted oak, a symbol of a lost past as
far back as Virgil, were seen as evidence of an earlier forested state,
whereas the barrows and monoliths were evidence of greater population
density and more intensive land use. The barrows with their burial goods
of sword and shield, and the runestones with their inscriptions, were
seen as landscape evidence of a distant past when Denmark was a world
power with, Blicher believed, a free, armed, farmer populace which
preserved the peace of the countryside. At the time Blicher began his
literary career the nation was in a social and financial crisis following
upon a disastrous war which had resulted in the loss of Norway, a legal
part of Denmark since the Middle Ages. This was only the last in a
series of wars, beginning in the seventeenth century, which had resulted
in considerable territorial loss. Jutland, in particular, had borne the
brunt of the ravages of these wars and this was seen as causing the
neglect of the land and the spread of the heath. It was thus little wonder
that Blicher found the pastoral convention appealing because of its
emphasis upon the peaceful life of the countryside of a more ideal
past.

The true thrust of the pastoral, of course, is not a nostalgia for a lost past, but the creation of a vision of another reality which breaks the monopoly of established reality and becomes the vehicle for a new consciousness of what things can be. Blicher, therefore, though he does dwell on the past, remonstrates with both himself and the reader for doing so. Thus, after eulogising the lives of 'our forefathers', in the poem quoted above, he then asks:

> Eye, why do you stare!
> At the shadows of ancient times?
> Does your happiness live alone,
> In departed years?

As a vision of the future he then presents an image of rural peace amongst the stormy wilderness of the heath. This is an area of the so-called *Al Heath* which, with state support, had been brought under cultivation and planted with forest. The peaceful, restorative heroism of this agricultural activity compares poorly with destructive military adventures. 'Are the *Al Heath*'s heather thatched houses,' he asks,

> Not just as beautiful, as the castle's
> smoking ruins?
> Shall thistles and thorns delight the
> ruler's heart,
> More than rich grain fields? [37]

The idea of the *peaceful* countryside is meant literally in the pastoral convention. It was a life of peace as opposed to the warlike activities instigated by the aggressive life of the town. Farming is nevertheless an heroic endeavour, worthy of the respect of society. The peaceful heroism of common rural life was a frequent theme in Blicher's authorship, and it certainly was appropriate to his time, but it can equally be found in places and times as far distant as Henry David Thoreau's Walden Pond and Virgil's Rome, where fields lie 'unkempt' because war had denied 'due honour to the plough'.[38] The following comment by L. P. Wilkinson on Virgil applies generally to the pastoral convention as a whole:

> Virgil was a man of peace, and the one insistent political motif in the *Georgics* is that swords should be beaten into ploughshares . . . Farming offered an outlet for the pugnacious and domineering

instincts of man, and Virgil depicts it as a constant physical battle
with iron weapons against weeds and overgrowth, in which the victor
'lords it over the fields' — *imperat arvis*.[39]

Andersen's comparison of the shape of Jutland with that of a
runestone, lying between the North Sea and the Skagerak, the past
'spoken by its graves', the future 'unfolding its power', is quite
understandable in the light of the literary landscape of the heath created
by Blicher. Poetry such as that of Blicher and Andersen could help
'break the monopoly of established reality' and create the basis for a
new consciousness — the sort of consciousness which leads to a new
perception, which leads people to 'come' and 'look it over' themselves,
as Andersen wrote. Poetry alone, however, could hardly challenge the
position of established agronomic interests.

One person who did come and see for himself, and who appreciated
both the value of poetry and the need for practical measures, was an
army road engineer named Enrico Dalgas (1828—94). As an engineer,
employed in the area, he was impressed by the way cultivation followed
infrastructural improvement. The roads on the heath were notoriously
poor, thus hindering market agriculture and favouring more extensive
pastoral forms. There were many examples, such as Blicher's from the
Al Heath, of successful local initiative in cultivating and afforesting the
heath, and there was local interest in continuing this effort. It was no
doubt his own confrontation with this interest and success that led
Andersen to write his optimistic poem. Dalgas saw, however, that local
initiative was not enough. Infrastructural improvements, the creation of
artificial floating meadows, the planting of shelter belts and afforestation, all would require a capital investment exceeding local resources if
it were to be effective. National private and state capital, on the other
hand, would not be forthcoming unless a sceptical public were
convinced that the heath could and should be reclaimed. Dalgas proved
to be a person who both had the vision to conceive of a workable
concrete programme and to help create a new perception of the area. It
was Dalgas who took the initiative in the establishment of the Danish
Heath Society in 1866, and who became its leading figure.

Heath Transformation

The formation of the Heath Society must be seen against the background
of the national trauma of the Danish defeat at the hands of Bismarck's
Germany in 1864 and the consequent loss of the provinces of Slesvig
and Holstein. The nation was once again in very much the same

situation as when Blicher perceived the seeds of regeneration in the
landscape of the heath. This literary landscape could hardly have been
more germane than at this time. Dalgas' first, and most critical, effort as
a publicist for what he called 'the heath cause' was a small book,
published in 1867–8, entitled *Geographical Pictures from the Heath*.
It begins with a lengthy heathscape description by Blicher which
Dalgas declares is 'not poetry' but the given reality of the heath as it
still largely was, but as it need not remain.[40]

Dalgas treated this literary landscape quite literally. For him, the
past truly was spoken by Jutland's graves, where the future unfolded
its power. He expended considerable effort in researching the historical
geographical evidence which would show that the heath had once been
more densely populated, intensively cultivated and forested. He based
his findings upon historical record, legend, place-name evidence and
archaeological finds, particularly, of course, the location of the
barrows! With the aid of choropleth maps correlating this evidence, he
was able to present a cogent argument that the heath was not, in fact,
fixed and defined by nature, but was, in the final analysis, the result of
social causes. The destruction of the forest, he maintained, had exposed
the fields to wind erosion and a consequent loss of fertility which led
to their abandonment and the subsequent spread of the heath. The
reforestation of the heath would, by the same logic, contribute to the
restoration of favourable conditions for agriculture.[41] Not only could
society bring about this restoration of the landscape of a better past, it
had the moral obligation to do so! The restoration of this landscape
was tied not only to the regeneration of Jutland, but to the regeneration of Danish society as a whole. The Heath Society's motto thus
became 'What is lost without, shall be gained within' – a phrase
known to virtually every Dane.

The external military loss was to be compensated for by an internal
victory which, though peaceful, was no less heroic. Dalgas thus made
liberal use of martial metaphor. The belts of new planted forest were to
become 'barricades' against the west wind enemy, and the Heath
Society itself was described as a 'phalanx' which has 'dug the trenches
in pursuance of that objective, which must be conquered'. Given the
historical situation, it was effective rhetoric, and, as has been seen, these
military allusions still persist in modern geographical descriptions of the
heath's transformation.[42]

Andersen's statement that 'in a few years, the heath a grainfield will
be' proved, of course, to be an accurate depiction of reality as it was to
become. The needed government and private capital investments were

made in the name of the patriotic 'heath cause', and the area
experienced the remarkable development which is recorded in standard
geographical descriptions. Opponents complained that their arguments
were drowned out by the patriotic 'Chinese gong' of the Heath Society,
but there was little they could do. They were not necessarily proved
wrong. It is possible that the capital might have been better invested
elsewhere, it is certainly true that some of the reclaimed land has since
reverted to more extensive uses, and that some afforestation was of
questionable value.[43] Still, there is no question that the life and
landscape of the area was changed drastically, as today's modern farms,
factories and growing cities amply testify.

The most eloquent testimony to the psychological importance of the
literary landscape of the heath is to be found in a park established in
the 1950s as a memorial to the undertaking. In an extensive portion of
the *Al Heath*, which has been carefully preserved (as has much of the
remaining heath), a hollow has been lined with modern runestones
commemorating the peaceful heroes who transformed the landscape.
Dalgas is there, of course, so is Blicher. Blicher's presence is
manifested, not just by a name on a stone, but by the whole,
seemingly paradoxical, idea of preserving the very landscape against
which the campaign had been directed. In the final analysis it was truly
a *heath* movement led by a *Heath* Society, in which it was this
landscape, as perceived through the form of literature, which spoke of
the past. It was also in this landscape that the future indeed did 'unfold
its power'. Without the apparently fixed and definable reality of the
heath, which nevertheless proved to be a deceptive reality, the 'heath
cause' loses its significance, and the measure of its achievement is lost.
As Marcuse writes:

> The utopia in great art is never the simple negation of the reality
> principle but its transcending preservation (Aufhebung) in which
> past and present cast their shadow on fulfillment. The authentic
> utopia is grounded in recollection.[44]

Conclusion

The view of art presented here may seem to differ from the approach,
prevalent in geography, which focuses upon the 'heightened perception'
of the individual artist. Nevertheless, the ability of the individual

artist to depict another, 'more genuine', reality, does depend upon perceptive insight into established reality. This is because, in Marcuse's words,

> The world intended in art is never and nowhere merely the given world of everyday reality, but neither is it a world of mere fantasy, illusion, and so on. It contains nothing that does not also exist in the given reality, the actions, thoughts, feelings, and dreams of men and women, their potentialities and those of nature.[45]

The literary landscape of the heath was thus not simply a figment of Blicher's imagination. Dalgas was able to give it credible scientific justification because it was well grounded in fact, as well as in artistic convention. In addition to his efforts in the realm of *belles lettres*, Blicher was also active in the area of agronomics and as the author of serious regional monographic studies. Blicher knew the area, and its probable past, better than most. The ability to create a credible vision of a more genuine reality than that which is given requires, if anything, more insight into the workings of things than a simple descriptive replication of the status quo. It is the sort of insight which would be of value to the geographic planner.

In the fullest sense, reality is both what things are and what they can be, and our experience of that reality is determined both by its actuality and what we believe it can become. It is nevertheless valuable to distinguish between the *is* and *ought* of literary reality, and to distinguish the role of the individual artist from the context of the socially established artistic form within which the artist operates to a greater or lesser degree. Once this is done, however, it becomes clear that the perspectives taken, for example, by Gilbert and Seamon are not necessarily in conflict with the view taken here. When Gilbert finds 'reality faithfully shown' in literature as compared to the 'dim twilight of modern geographical jargon', perhaps he is really finding a more genuine reality than the established reality of the narrow economic concerns of technocratic planners. Likewise, one might ask whether it really is the sensitive individual artist whose experiences, as articulated in literature, provide him with an alternative awareness of his *own* geographical situation. Perhaps it is, in fact, the creation of an alternative reality, in the form of art, which enables him to become aware of patterns in his *own* experience that he had not known before.

Notes

I would like to thank Yi-Fu Tuan for his critique. By discerning the weak points in my argumentation he has helped clarify my thought. This does not make him responsible for the final content of the manuscript.

I would also like to thank the Danish State Research Council for providing financial support, and Niels Hansen for cartographic assistance.

1. D. C. D. Pocock, 'The Novelist and the North', *Occasional Publications*, New Series, no. 12 (Department of Geography, University of Durham, 1978), p. 1.

2. C. L. Salter and W. J. Lloyd, 'Landscape in Literature', *Resource Papers for College Geography* (Association of American Geographers, Washington DC), no. 76–3 (1977), p. 3.

3. Ibid.

4. E. W. Gilbert, 'The Idea of the Region', *Geography*, vol. 45 (July 1960), p. 167.

5. David Seamon, 'The Phenomenological Investigation of Imaginative Literature' in G. T. Moore and R. G. Golledge (eds.), *Environmental Knowing: Theories, Research and Methods* (Dowden, Hutchinson & Ross, Stroudsburg, Pa., 1976), p. 290.

6. Northrop Frye, *Anatomy of Criticism: Four Essays* (Princeton University Press, Princeton, NJ, 1957), p. 99.

7. Susanne K. Langer, *Feeling and Form: A Theory of Art* (Scribners, New York, 1953), pp. 59–60.

8. Herbert Marcuse, *The Aesthetic Dimension* (Beacon Press, Boston, 1978), p. 8.

9. Herbert Marcuse, 'Art as Form of Reality', *New Left Review*, no. 74 (July–August 1972), p. 52.

10. Yi-Fu Tuan, 'Literature and Geography: Implications for Geographical Research' in D. Ley and M. S. Samuels (eds.), *Humanistic Geography: Prospects and Problems* (Croom Helm, London, 1978), p. 205.

11. Langer, *Feeling*, p. 49.

12. Marcuse, *Aesthetic*, pp. 41 and 49.

13. Ibid., p. 54.

14. Frye, *Anatomy*, p. 184.

15. L. Marx, *The Machine in the Garden* (Oxford University Press, London, 1964); R. Williams, *The Country and the City* (Oxford University Press, New York, 1973); 'Literature and Rural Society', *The Listener*, 16 November 1967, pp. 630–2. See Yi-Fu Tuan, *Topophilia* (Prentice-Hall, Englewood Cliffs, NJ, 1974), pp. 102–12; 'Book review: *The Country and the City*', *Landscape*, no. 22 (1978), pp. 19–20.

16. Williams, 'Literature', pp. 630–1.

17. Ibid., p. 631.

18. Jane Zaring, 'The Romantic Face of Wales', *Annals of the Association of American Geographers*, no. 67 (September 1977), pp. 397–418.

19. See William Wordsworth, *A Guide to the District of the Lakes in the North of England*, 1810 (Greenwood Press, New York, 1968). See especially W. M. Merchant's introduction.

20. Marcuse, *Aesthetic*, p. 54.

21. Frye, *Anatomy*, pp. 169–70.

22. Pocock, 'Novelist'; see also his 'The Novelist's Image of the North', *Transactions of the Institute of British Geographers*, vol. 4 (1979), pp. 62–76.

23. Pocock, 'Novelist', p. 22. Quoted from D. H. Lawrence, *Lady Chatterley's Lover*, 1928 (Heinemann, London, 1960), p. 139.

24. Pocock, 'Novelist's Image', pp. 72–3.

25. Pocock, 'Novelist', p. 32. Quoted from A. Bennett, *Anna of the Five Towns*, 1902 (Penguin, Harmondsworth, 1954), p. 25.

26. Williams, *Country*, pp. 13–34.

27. Frye, *Anatomy*, p. 99; *A Natural Perspective* (Columbia University Press, New York, 1965), pp. 61–2.

28. Oliver Goldsmith, *Collected Works of Oliver Goldsmith*, vol. 4, edited by Arthur Friedman (Clarendon Press, Oxford, 1966), pp. 287–304. See Introduction, pp. 273–81.

29. Harry Thorpe, 'A Special Case of Heath Reclamation in the Alheden District of Jutland, 1700–1955', *Transactions and Papers, Institute of British Geographers*, no. 23 (1957), p. 87.

30. Brian Fullerton and Allan F. Williams, *Scandinavia* (Chatto & Windus, London, 1972), pp. 111–12.

31. Kenneth Olwig, 'The Morphology of a Symbolic Landscape: The Transformation of Denmark's Jutland Heaths Circa 1750–1950', unpublished PhD dissertation, University of Minnesota, 1977, pp. 269–75.

32. Ibid., pp. 338–78.

33. Hans Christian Andersen, 'Jylland Mellem Tvende Have', *Folkehøjskolens Sangbog*, 15th edn (Foreningen for Højskoler og Landbrugsskoler, Odense, 1964), no. 530, pp. 714–15. All translations by the present author.

34. Olwig, 'Morphology', pp. 277–81; see also his 'Place, Society and the Individual in the Authorship of St. St. Blicher' in Felix Nørgaard (ed.), *Omkring Blicher 1974* (Gyldendal, Copenhagen, 1974).

35. Steen Steensen Blicher, 'Jyllandsrejse i Sex Døgn' in his *Samlede Skrifter*, vol. 4, edited by Jeppe Aakjaer and Henrik Ussing (Gyldendal, Copenhagen, 1920), p. 214.

36. Steen Steensen Blicher, 'Hiemvee' in *Skrifter*, vol. 3, p. 137.

37. Blicher, 'Jyllandsrejse', pp. 215 and 223–4.

38. Virgil, *Georgics*, 1: 596–9.

39. L. P. Wilkinson, 'The Intention of Virgil's Georgics', *Greece and Rome*, no. 19 (January 1950), pp. 21–2.

40. E. Dalgas, *Geographiske Billeder fra Heden*, part 1 (Reitzel, Copenhagen, 1867), pp. 3–4. See also Olwig, 'Morphology', pp. 338–51.

41. Olwig, 'Morphology', pp. 338–51.

42. Dalgas, *Geographiske*, part 2 (1868), pp. 124–5.

43. See Kr. Marius Jensen, 'Opgivne og Tilplantede Landbrugsarealer i Jylland', *Atlas Over Danmark*, series 2, vol. 1 (Reitzel, Copenhagen, 1976).

44. Marcuse, *Aesthetic*, p. 73.

45. Ibid., p. 54.

4 CONSCIOUSNESS AND THE NOVEL: FACT OR FICTION IN THE WORKS OF D. H. LAWRENCE

Ian G. Cook

The humanist geographer and the novelist have much in common. Both seek to portray the activities of people within the context of a specific milieu, infusing their descriptions of people and places with a sensitivity born of a rich and varied experience of life and society. Both seek to engender in their audience a deep awareness and empathy concerning the situation of others and their *lebenswelt*. And both may be equally biased in the presentation of their world views, wittingly or unwittingly distorting and manipulating the material at their command. The geographic use and interpretation of the novel must, therefore, face the bias of the novel — its 'poetic licence' as it were — and attempt to develop a critical analysis which is fully cognisant of its defects and limitations. This analysis hinges upon the concept of *consciousness*, which in its various forms is central to our appreciation of the content of the novel and our evaluation of its usefulness in geographical studies. In this chapter the concept of consciousness will be briefly reviewed before being applied in the evaluation of the works, particularly the coalfield novels, of D. H. Lawrence.

The Concept of Consciousness

Within geography the concept of consciousness can be associated with the works of humanist geographers on the one hand or radical geographers on the other.[1] The former stress that consciousness is generated from within the individual, whereas the latter emphasise that consciousness is the product of external, social processes impinging upon the individual. Both approaches are of course concerned with the interaction between the individual and society, but humanist geographers consider consciousness to be the result of the individual's interpretation of the world, flowing outward to society, while radical geographers consider it to be the result of the individual's position in society, flowing from society inward to the individual. The humanist focuses upon the individual's experience of life, his or her values, attitudes and beliefs, the meanings attached to phenomena, and other 'subjective' factors, and studies consciousness via this route. In contrast, the radical begins by analysing the person's class position, the relationship between this class and the ruling class in society, and the susceptibility which this class exhibits to the absorption of the ideology disseminated by the ruling class. Thus the radical uses the term 'false consciousness' to describe the manner in which consciousness bears no relationship to external reality, being a product of the ruling-class manipulation of information flows, societal institutions and the like. In Peet's words, 'The system-supportive ideologies serve to blunt the effects of sharp events, provide diversionary explanations for their causes, and in a myriad other ways act to promote false consciousness.'[2]

Whichever of these views of consciousness is accepted will markedly influence the geographer's use of the novel. If it is accepted that consciousness arises essentially from within, then the novel will be used to enhance this consciousness, and the novel will be accepted as a means of enhancing the geographical imagination and sense of place of the geographer. Acceptance of the view that consciousness is primarily a social product leads alternatively to a more critical analysis of the novel, and indeed of the novelist also. In the latter perspective the novel cannot simply be appreciated as an independent work of art, but rather must be viewed in the context of the novelist's objectives in writing the novel, the influences which have given rise to the particular form and content which it presents, and the social context in which the novelist is operating. Specifically, the novel must be scrutinised lest it be written in such a way that it promotes the false consciousness noted previously.

In scrutinising the novel for evidence that it promotes false consciousness one is essentially searching for evidence of class bias in its writing, for the consciousness is in the latter perspective clearly linked to class. This concept of class consciousness can in turn be related to Parkin's ideas concerning variation in *value systems*, value systems being closely linked to consciousness of reality, underpinning this concept. Thus Parkin identifies in modern Western society three major value systems, or *meaning systems*, which are:

1. The *dominant* value system, the social source of which is the major institutional order. This is a moral framework which promotes the endorsement of existing inequality; among the subordinate class this leads to a definition of the rewards structure in either *deferential* or *aspirational* terms.
2. The *subordinate* value system, the social source of which is the local working-class community. This is a moral framework

which promotes *accommodative* responses to the facts of
inequality.

3. The *radical* value system, the source of which is the mass political
   party based on the working class. This is a moral framework
   which promotes an *oppositional* interpretation of class
   inequalities.[3]

These value systems are related to consciousness in that 'the subordinate value system restricts man's consciousness to the immediacy
of a localised setting; and the dominant value system encourages
consciousness of a national identity; but the radical value system
promotes the consciousness of class'.[4]

Use of the novel by the radical geographer therefore requires him or
her to discover whether the novel is system-supportive, upholding the
dominant value system, accommodative in that it largely ignores
matters of class, replacing these by concern for local matters, or
whether it is radical, providing a challenging critique of the
dominant ideology, a critique which is based on class. If such a critique
is presented in the novel, then it can be said to promote *counter-
consciousness*,[5] or a radical awareness of societal inequalities and the
reasons for these.

The choice between these alternative approaches is a difficult one,
demanding knowledge of the idealist and materialist philosophies in
which they are, respectively, grounded.[6] Synthesis may be possible for
the relationship between them may well be dialectical, but this is too
complex an issue to be entered upon here, and a simpler aid to the
choice must be sought. To this end it may be useful to consider the
novel as a form of the mass medium, and consider the way in which it
presents information to the reader.

The Novel as a Medium

The novel can be considered as a medium insofar as it presents second-
hand information to the reader. As with other forms of media, the
novel thus allows us to escape the bounds of our own immediate
experience of the world, presenting us with a rich source of 'data'
which complements and supplements the first-hand information which
we receive. The novel thus acts as a 'communication channel' between
some 'reality' and our personal images of reality. In a similar fashion
radio, television and other forms of media also act as channels of

communication, channels which are subject to selection and distortion of information, as Galtung and Ruge make clear in their study of the structure of foreign news.[7] Their model of a chain of communication from world events to personal image may be adapted to the novel, and is shown in Figure 4.1. In this reformulation the novelist perceives 'reality', using perception in the broadest sense, and the novelist's images then develop from this process of perception, being selected and distorted en route.[8] Further selection and distortion then occurs in the process of the writing of the novel and the image, or images, which the novel presents are influenced by such factors as the novelist's views of what the public wants, his publisher's views of this, editing by the latter and so on, while the act of writing and rewriting of the novel will also modify its content.[9] And finally, the reader himself or herself will introduce a further element of selection and distortion via the process of his or her perception of the novel and evaluation of it.

The selection and distortion noted above reflect a variety of factors but these can be simplified as suggested in Figure 4.2, with distortion being regarded as the product of the situation, or context, in which the novelist and the reader find themselves. 'Situation' sums up the psychological and social influences upon them, those that reflect environmental and spatio-temporal location, and the other manifold influences upon perception,[10] and serves as a term which reminds us forcefully that the novel cannot be viewed in isolation from the novelist. Further, 'situation' suggests the wider world in which the novelist operates, and lends support to the view of consciousness

Figure 4.1: Chain of Communication of the Novel

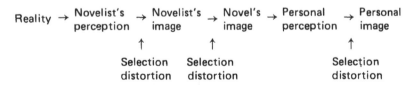

Figure 4.2: Simplified Chain of Communication of the Novel

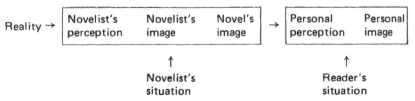

expressed by radical geographers, and to the need to scrutinise the nature of the novelist as well as the novel, while the reader's situation must also be studied. In proceeding with an evaluation of Lawrence's works, therefore, it will be the radical approach which will loom largest in the critique which is undertaken; therefore Lawrence himself must be studied, as well as his major works. The next section will, accordingly, present a brief outline of Lawrence's background in order that his writings may be better understood.

Lawrence's Background

David Herbert Lawrence was born on 11 September 1885 in 8a Victoria Street, Eastwood, a small town in the Nottinghamshire—Derbyshire coalfield area. The town[11] lies upon a hill overlooking the Erewash valley, along which ran three canals and two major railway lines in Lawrence's day, being used to carry coal to the Nottingham and London areas respectively. During the time of his upbringing industrial development had only just begun to spread over the landscape: mines were an 'accident' and 'Robin Hood and his merry men were not far away'.[12] As a result, Lawrence became acutely aware of the increasing encroachment of industry and the despoliation which it heralded. This awareness would have been heightened by his acute sensitivity, in part engendered by a severe illness during his childhood, by the fact that he 'was most receptive to the beauty of nature', and by the fact also that his family shared this receptiveness. His mother loved trees, plants and flowers, while even his miner father 'was very good on wild life . . . knew the names of the birds and animals . . . [and] had a reverence for wild life'.[13]

Another major influence upon Lawrence's attitude towards industry seems to have been the association of industry with the negative aspects of his father's life — strife, brutality, dirt, grime, poverty — in contrast with the association of his mother with nature, with the emphasis upon beauty, love, affection, cleanliness and other positive features. That which the parents symbolised to Lawrence, at least in his earlier years, were thus transferred to industry and nature respectively. These associations reflect the turbulent relationship between Lawrence's parents and the fact that for many years Lawrence hated his father.

This hate for his father requires further examination, for it seems to have had further ramifications than that noted above. Lawrence's father was a miner, and was thus a member of the 'lower classes' of Eastwood society.[14] His mother, conversely, had been a schoolteacher prior to her marriage and had thus 'married below her station'. Moreover, she was a congregationalist who had a rather austere and puritanical view of life and was 'of the temperance persuasion', whereas Lawrence's father 'was a robust, black bearded little man who sometimes stopped at the pub on the way home and spent part of his wages'.[15] His mother would belittle him for this in front of the children and a fight would thus be provoked, often resulting in violence. These fights are described in the opening chapters of *Sons and Lovers*, an essentially autobiographical novel.[16] As the novel relates, they had a great impact upon the children, who began to abhor their father, taking the mother's part in these quarrels. Gradually the parents became more and more estranged, the mother transferred her affections to the children, especially to Lawrence during his illness, and the father was shut out from family life. This of course exacerbated the situation, for he was driven all the more to the public house to seek solace with his friends.

Faced with such a home life, it seems highly probable that Lawrence gradually transferred his feelings for his parents to that which they symbolised, industry and nature respectively. There was the contrast between the unspoilt countryside around Eastwood compared to the spoiling effects of the spread of industry and of the town itself; and there was the contrast between the deep, dark, sensuous nature of his father compared to the more refined intellectual nature of his mother. Combined with his own imaginative view of the world, there developed within him a great awareness of place. Lawrence was later to record that

> Every continent has its own great spirit of place. Every people is polarized in some particular locality, which is home, the heartland. Different places on the face of the earth have different vital effluence, different vibration, different chemical exhalation, different polarity with different stars: call it what you like. But the spirit of place is a great reality.[17]

From the age of fourteen his awareness began to extend beyond the immediate environment into other areas as a result of his prolific reading. This in turn was stimulated by his mother, and then by his relationship with Jessie Chambers[18] of Haggs Farm, near Eastwood, before he became a teacher.[19] T. S. Eliot has criticised Lawrence for being 'an ignorant man, in the sense that he was unaware of how much he did not know',[20] but although Lawrence can be criticised on several grounds, this particular criticism appears unjustified. Leavis and others have defended Lawrence from this charge, and the defence seems just.[21] In 1904, for example, Lawrence came first in the King's Scholarship in England and Wales, and although poverty at first prevented him from utilising the grant which this provided, in 1906 he was at last able to begin a two-year teaching course at Nottingham University College, as it then was. There he gained distinction in many of his subjects, and read widely in many disciplines. He was thus far from ignorant, in any sense of the term.

Lawrence then took up employment as a teacher in Croydon where he met many influential and well-known people in the avant-garde set, and had a number of generally unsatisfactory relationships before his elopement with Frieda von Richtofen, the wife of Professor Weekley of Nottingham University College. Lawrence's relationship with his mother may have been Oedipal, and was certainly inhibiting even after her death in 1910, but Frieda helped him to unshackle himself from these ties without replacing them with inhibitory ties of her own. Instead their's was a 'conflict of love and hate',[22] a fierce clash of wills which provided a milieu which further heightened his sensitivity. Frieda also provided the stimulus to his travels, for he initially eloped with her to Germany in 1912.

Following this journey, Lawrence's first trip to the Continent, they spent most of the next eighteen years in Italy or the South of France. They were entombed in England during the First World War, however, suffering greatly because of Lawrence's outspoken anti-war views, his frequent illness, publication of *The Rainbow* in 1915 (banned, ostensibly for obscenity, but more probably because of its anti-war views) and by the fact that Frieda was German. After the war they were at last allowed to leave England for Italy, and before Lawrence's untimely death in 1930 they were able to visit Ceylon, Australia and North America, these places providing the setting and ideas for many of his later works.[23] The travels were also of importance in that they presented living environments with which he could compare his native Midlands, giving a perspective upon them. This point is important, for as Bentley has noted, the best regional novels have been written only after the author has had the opportunity to travel widely in order to obtain this necessary perspective.[24]

The changing perspective which travel and experience brings was reflected not only in changing views of the Midlands, but also in changing attitudes towards his father. Thus in 1922 he could inform a friend that 'he had been unjust to his father in *Sons and Lovers*';

accepting his mother's view of him, he had not realised 'his unquench-
able fire and relish for living'. When his father came in after drinking
with his friends, she never allowed him to go quickly and quietly to
bed, but reviled him while he told the children they had nothing to
fear.[25] It was noted previously that Lawrence's views of his parents
could be linked to his views of industry and nature respectively, but
this analysis must clearly now be modified in the light of these changing
views. Thus, as Sanders notes, Lawrence's father can be, and was,
associated with nature, for he was the vehicle for sensuality, feeling,
emotion and deep undercurrents of sexuality and the rawness of nature.
In contrast, Lawrence's mother, with her puritanical views, began to be
increasingly associated with order, organisation and 'culture' which
could be expressed in the increasing control over human life which
industry brought.[26]

These contrasting views struggled for dominance within Lawrence
and were expressed, as we shall see, in changing views of nature and
industry (interpreted in the widest sense), with the former sometimes
being described in almost antiseptic terms and sometimes in deep, full-
blooded ones, while industry could either be identified with deep, dark
forces, particularly where the miners themselves were portrayed, or with
order and control where the system was described. And finally, but no
less appositely, it must be noted that this ambivalence of views of the
world has an important class dimension in that the fact that Lawrence's
parents were in effect from different class backgrounds is interwoven
with these views, with the result that Lawrence sometimes expresses
class bias in his portrayal of the people in his novels. Indeed, Sanders
has argued that Lawrence, because of his background, is *déclassé* and
alienated from it,[27] and this point must be borne in mind when his
novels are examined.

I have dwelt at some length upon Lawrence's background, for it
seems to me to be of much importance in coming to an understanding
of his writings. Even so, much detail has been omitted and readers are
recommended to the many biographies of Lawrence which have been
published for more information about him.[28] The essay will now turn
to the coalfield novels themselves in order to discover how this
background finds expression in Lawrence's writings. This can be done
through study of descriptions of 'nature' and of 'industry' respectively.

Lawrence and Nature

Lawrence loved the countryside around Eastwood, particularly the area
to the east-north-east, near Haggs Farm, the home of Jessie Chambers.
This he describes as 'the country of my heart';[29] it is captured so

beautifully in his first novel, *The White Peacock*, and in Part Two of *Sons and Lovers*. In the latter we read

> Willey Farm [i.e. Haggs Farm] he loved passionately . . . the little pokey kitchen, where men's boots tramped and the dog slept with one eye open for fear of being trodden on . . . [the] long, low parlour, with its atmosphere of romance, its flowers, its books, its high rosewood piano. He loved the gardens and the buildings that stood with their scarlet roofs on the naked edge of the fields, crept towards the wood as if for cosiness, the wild country scooping down a valley and up the uncultured hills of the other side. Only to be there was an exhilaration and a joy to him.[30]

Nature is portrayed at all scales in these novels, from detailed descriptions of the flora and fauna to wider perspectives upon the valley within which the farm lay. At times Lawrence's knowledge and perception is quite remarkable in its detail. Take, for example, the following extract from *The White Peacock*:

> So we went along the hurrying brook, which fell over little cascades in its haste, never looking once at the primroses that were glimmering all along its banks. We turned aside, and climbed the hill through the woods. Velvety green sprigs of dog-mercury were scattered on the red soil. We came to the top of a slope where the wood thinned . . . I found that I was walking, in the first shades of twilight, over clumps of snowdrops. The hazels were thin, and only here and there an oak tree uprose. All the ground was white with snowdrops, like drops of manna scattered over the red earth, on the grey-green cluster of leaves. There was a deep little dell, sharp sloping like a cup, and a white sprinkling of flowers all the way down, with white flowers showing pale among the first impouring of shadow at the bottom. The earth was red and warm, pricked with the dark, succulent green of bluebell sheaths, and embroidered with grey-green clusters of spears, and many white flowerlets . . . it seemed like a holy communion of pure wild things, numberless, frail, and folded meekly in the evening light.[31]

A beautiful scene. And yet there lies within the description an element of neutrality, a lack of passion, for this was a novel which was written for the approval of his mother. Even in *The White Peacock*, however, there are scenes of greater power, which play upon all the senses so that we are transported into the place, seeing it, touching it, feeling it, experiencing it in all its facets. By such means is 'spirit of place' conveyed.

Places then are clearly much more than physical entities. As Sinzelle notes, 'Places are not abstract entities which exist in the absolute, they are inhabited by people who give them a sense.'[32] It is pertinent, therefore, to consider the rural people to whom Lawrence refers. One to be particularly noticed is George Saxton, 'a young farmer, stoutly built, brown eyed, with a naturally fair skin burned dark and freckled in patches'. His character is exposed by the way in which he 'looked . . . with lazy curiosity', 'with a lazy indulgent smile', that he spoke in a 'leisurely fashion' and read 'indolently', being 'so content with his novel and his moustache'.[33] Later in the novel he becomes a drunkard, this being written to show the evils of drink, for, as we have noted above, this novel was written for the approval of Lawrence's mother, and the immaturity of the young writer is evident in this and other examples from it.

The young writer matured, however, and his art reached a peak in *The Rainbow* and *Women in Love*,[34] widely regarded as being his best novels. *The Rainbow* is particularly relevant here for it was set in and around the lovely village of Cossall, several miles to the south of Eastwood. The opening pages have become famous and deserve to be quoted at length, for they describe the bonds between man and land so effectively:

> The Brangwens had lived for generations on the Marsh Farm, in the meadows where the Erewash twisted sluggishly through alder trees, separating Derbyshire from Nottinghamshire. Two miles away, a church tower stood on a hill, the houses of the little country town climbing assiduously up to it . . . so that [when a Brangwen looked up from his work] he was aware of something standing above him and beyond him in the distance. There was a look in the eyes of the Brangwens as if they were expecting something unknown, about which they were eager. They had that air of readiness for what would come to them, a kind of surety, an expectancy, the look of an inheritor.
>
> They were fresh, blond, slow-speaking people, revealing themselves plainly, but slowly, so that one could watch the change in their eyes from laughter to a hard blue-staring anger; through all the irresolute stages of the sky when the weather is changing. . .
>
> Living on rich land . . . they came and went without fear of

necessity . . . Neither were they thriftless . . . They knew the intercourse between heaven and earth, sunshine drawn into the breast and bowels, the rain sucked up in the daytime . . . feeling the pulse and body of the soil, that opened to their furrow for the grain, and became smooth and supple after their ploughing, and clung to their feet with a weight that pulled like desire, lying hard and unresponsive when the crops were to be shorn away . . . They took the udder of the cows, the cows yielded milk and pulsed against the hands of the men, the pulse of the blood of the teats of the cows beat into the pulse of the hands of the men. They mounted their horses, and held life between the grip of their knees, they harnessed the horses at the wagon, and, with hand on the bridle-rings, drew the heaving of the horses after their will.[35]

This passage, so rich in tone and infused with a deep sensuous feeling and sexual potency, graphically and poetically illustrates the rhythm and beat of rural life, and the blood-intimacy between the Brangwens and nature.

Thus is nature presented by Lawrence, but before attempting an evaluation of this, let us turn to his portrayal of the urban-industrial world, a world of marked contrast to the above.

Lawrence and Industry

Lawrence grew to detest industrialisation and associated urbanisation with their impact upon society. In his early childhood this was not the case, however, if this example from *Sons and Lovers* is a guide. In it Clara, his girlfriend at the time, comments that it is a pity that Moorgreen Colliery is there in an otherwise pretty area, and Paul replies:

Do you think so? . . . You see I am so used to it I should miss it. No; and I like the pits here and there. I like the rows of trucks, and the headstocks, and the steam in the daytime, and the lights at night. When I was a boy, I always thought a pillar of cloud by day and a pillar of fire by night was a pit, with its steam, and its lights, and the burning bank – and I thought the Lord was always at the pit top.[36]

The pits nevertheless spewed out ash, which had a considerable impact upon surrounding people and places. Thus Lawrence could describe the houses of the Bottoms (i.e. the Breach) in Bestwood (i.e.

Eastwood) as being 'substantial and very decent', although the actual living conditions were 'quite unsavoury'.[37] But despite his criticisms, life in the Bottoms none the less seems bearable, which is not the case for Wiggiston in *The Rainbow*:

> Wiggiston was only seven years old. It had been a hamlet of eleven houses on the edge of heathy, half-agricultural country. Then the great seam of coal had been opened. In a year Wiggiston appeared, a great mass of pinkish rows of thin, unreal dwellings of five rooms each. The streets were like visions of pure ugliness; a grey-black macadamized road, asphalt causeways, held in between a flat succession of wall, window, and door, a new-brick channel that began nowhere, and ended nowhere. . .
>
> The place had the strange desolation of a ruin. Colliers hanging about in gangs and groups, or passing along the asphalt pavements heavily to work, seemed not like living people, but like spectres. The rigidity of the blank streets, the homogeneous amorphous sterility of the whole suggested death rather than life. There was no meeting place, no centre, no artery, no organic formation. There it lay, like the new foundations of a red-brick confusion rapidly spreading, like a skin-disease.[38]

In this passage we are near the zenith of Lawrence's evocative condemnation of the impact of industrialisation, and are light years away from the world of the Brangwens in their rural setting. It is in *Lady Chatterley's Lover* that this evocation reaches its peak, in passages written during some of the darkest years in British history. Towns such as Teversal, a small town several miles to the north-east of Eastwood, plunged to the nadir of their development:

> The car ploughed uphill through the long squalid straggle of Tevershall [i.e. Teversal], the blackened brick dwellings, the black slate roofs glistening their sharp edges, the mud black with coaldust, the pavements wet and black. It was as if dismalness had soaked through and through everything. The utter negation of natural beauty, the utter negation of the gladness of life, the utter absence of the instinct for shapely beauty which every bird and beast has, the utter death of the human intuitive faculty was appalling . . . all went by ugly, ugly, ugly. . .
>
> Tevershall! That was Tevershall! Merrie England! Shakespeare's England! No, but the England of today . . . producing a new race of

mankind, over-conscious in the money and social and political side, on the spontaneous, intuitive side dead, but dead. . .

   She felt again in a wave of terror the grey, gritty hopelessness of it all. With such creatures for the industrial masses, and the upper classes as she knew them, there was no hope, no hope any more.[39]

In his description of Tevershall his fulminations against industriali-sation reach a peak. But what of the social system? This is given in the following passage, from *The Lost Girl*, published in 1920:

> Take a mining townlet like Woodhouse [i.e. Eastwood] with a population of ten thousand people, and three generations behind it . . . The old 'County' has fled from the sight of so much disembowelled coal, to flourish on mineral rights in regions still idyllic. Remains one great and inaccessible magnate, the local coal-owner; three generations old, and clambering on the bottom step of the 'County', kicking off the mass below. Rule him out.[40]

The 'County' is of course the old establishment, or aristocracy of the county, which has escaped from the dirt largely responsible for their wealth to 'regions still idyllic'. In these few words we see the impact of coalmining upon the landscape, and also the nature of the new pseudo-aristocracy, the social climber who is seeking to join the establishment and gain superior status. Indeed, 'Rule him out.' But what of the rest of society? Lawrence continues:

> Here we are then; a vast substratum of colliers; a thick sprinkling of tradespeople intermingled with small employers of labour and diversified by elementary schoolmasters and nonconformist clergy; a higher layer of bank-managers, rich millers and well-to-do ironmasters, episcopal clergy and the managers of collieries: then the rich and sticky cherry of the local coal-owner glistening over all.
>   Such is the complicated social system of a small industrial town in the Midlands of England, in this year of grace 1920.[41]

In these words the society is laid bare, like rock strata in a cutting. At the foot are the colliers and then come the shopkeepers such as James Houghton in *The Lost Girl* or the teachers such as Ursula in *The Rainbow*, people who fit rather uneasily into the system for they are neither of the colliers nor of the owners, although they have links with both. And so on up the social hierarchy to the coal-owner, such as

Gerald in *Women in Love* or Clifford in *Lady Chatterley's Lover.* Thus
the social structure of the town is represented, with its latent class
snobbery being exposed wittingly or unwittingly by the author. The
contrast between the miners and the coal-owner is particularly wide, as
illustrated by the following extract from *The Lost Girl*:

> Slaves of the underworld! She watched the swing of the grey colliers
> along the pavement with a new fascination, hypnotized by a new
> vision. Slaves – the underground trolls and iron-workers, magic,
> mischievous, and enslaved, of the ancient stories. But tall – the
> miners seemed to her to loom tall and grey, in their enslaved magic.
> Slaves who would cause the superimposed day-order to fall. Not
> because, individually, they wanted to. But because, collectively,
> something bubbled up in them, the force of darkness which had no
> master and no control. It would bubble and stir in them as earth-
> quakes stir the earth. It would be simply disastrous, because it had
> no master. There was no dark master in the world. The puerile
> world went on crying out for a new Jesus, another Saviour from the
> sky, another heavenly superman. When what was wanted was a Dark
> Master from the underworld.
>   So they streamed past her, home from work – grey from head to
> foot, distorted in shape, cramped, with curious faces that came out
> pallid from under their dirt. Their walk was heavy-footed and
> slurring, their bearing stiff and grotesque. . . The miners, the iron-
> workers, those who fashion the stuff of the underworld.[42]

This extract, which will be commented on below, may be compared
with the reference to Gerald as 'the God of the machine' in *Women in
Love*.[43] Although Gerald succeeds in changing the system, destroying
the last vestige of its humanity, he is in turn destroyed, by Gudrun.

The powerful imagery outlined in the extracts above is bewitching
in its persuasiveness. The aim of the present critique, however, is to
evaluate these scenes in the context of consciousness and their impact
upon it.

Evaluation: Consciousness and Lawrence

In the title to this chapter the question of fact or fiction in the works
of D. H. Lawrence was posed. In one sense this question is nonsense,
for his works *are* fiction and not fact, and yet it has been posed in this

way to act as a reminder that the novel contains degrees of truth, of verisimilitude, which is what the geographer is often most interested in. With regard to Lawrence's writings, their high autobiographical content lends itself fairly readily to this type of question, and to questioning of the consciousness which they arouse in the reader. Evaluation can only be tentative, and yet it must be attempted lest we direct our students to literature in the search for gold when all that can be panned is fool's gold. Evaluation in this case requires a combination of knowledge of Lawrence's background, and knowledge of critiques of his writings by scholars of literature, both of which have been summarised above, plus any geographical expertise and understanding which the present writer can bring to bear concerning the area about which Lawrence wrote.[44] In sum, these factors lead me to the following perspective upon his novels.

There is little doubt that Lawrence is at his strongest in the portrayal of place, whether it is the rural or the urban-industrial. In this I concur with Aldington, who wrote that 'it is always, or nearly always, the spirit of place which is evoked – the experience of that passionate sensibility which made Lawrence supreme in his time as the poet of the living world'.[45] This 'passionate sensibility' is probably the essence of Lawrence's genius, and reflects influences upon him which have already been described above. Although the descriptions of place sometimes appear to have an exaggerated sense of 'nature' on the one hand or of 'civilisation' on the other, this is part of the novelist's art, and is necessary to fully bring out the essence of place and of the changes which industry can wreak upon the landscape. Those quoted passages concerned with the physical fabric of place can thus be accepted at face value, and can be used to enhance our consciousness of the beauty of nature, the dangers of urbanisation and industrialisation, the sense of place of the East Midlands, and various other features of place which Lawrence presents so vividly.

Lawrence is also strong on the relationship between the individual and nature, and on the links between people and place. Here too we can readily accept his imagery and his attempts to elaborate upon the nexus at the subconscious or unconscious level, as in *The Rainbow*, for example.

However, where Lawrence is at his weakest, and hence at his most misleading, or even dangerous, is in his description, or lack of description, of the links between the individual and society. Often characters seem to exist in a society-less vacuum, with nothing being presented to show that their problems and difficulties reflect societal conditions. For Lawrence 'it is the change inside the individual which is my real concern. The great social change interests me and troubles me, but it is not my field.'[46] He thus shied away from an analysis of society in his writings, compared to Marx for example:

> Whereas for Marx the cause of suffering and dehumanization among workers was a particular form of society, a particular organization of industry, for Lawrence in his most extreme moments industry itself, society in whatever form, appears as the devil to be damned.[47]

As Sanders persuasively argues, Lawrence's background led to his social alienation and *déclassement*, with the result that he could not identify with *any* specific class in society. Class bias is therefore found in his writings, and this is marked whenever the 'masses' are described. Sanders suggests that Lawrence's dominant emotion with regards to the masses was one of fear:

> fear of being reabsorbed into the working class in which his father toiled, and from which he himself had emerged; fear of being identified with that class of people whom his mother had taught him to despise.[48]

This is not the complete answer, however. Phrases quoted above, such as 'creatures for the industrial masses' or 'Slaves who would cause the superimposed day-order to fall', not only exhibit class bias, but pander to the worst middle-class fears of miners, bolshevism and the like. In part, the imagery is mythical, as Kermode notes,[49] and in part it reflects his complex feelings for them, feelings which are often ambivalent and traceable to ambivalent feelings towards his father.

Lawrence's individualism, his alienation from his background, and his rejection of the upper classes also, led him to seek solutions to society's problems via unconventional paths. Indeed, Bertrand Russell accused Lawrence of fascism, and that Lawrence 'had developed the whole philosophy of fascism before the politicians had thought of it'.[50] Such criticism is the result of Lawrence's developing philosophy of the blood, of the 'blood-intimacy' of the Brangwens, for example, or the call for a 'Dark Master of the underworld' in *The Lost Girl*. Russell maintained that this philosophy was one that 'led straight to Auschwitz',[51] and so we must guard against the deep dark features and symbols in his novels lest we become prey to the false consciousness of society to which this philosophy can lead.

There is, then, a tendency for Lawrence's writings to lead to false consciousness as far as social processes are concerned. Lawrence believed 'that the idea of giving power to the hands of the working class is *wrong*'[52] and this viewpoint coloured his writings. Nevertheless, the complexity of his feelings for the miners specifically, and the working class in general, was such that he could occasionally give a truer picture of them. This is illustrated in *The First Lady Chatterley*,[53] as the first version of this novel is now known. In this first version, as Mellor notes, the character of the gamekeeper is more true to working-class values than in the third version, joining a Communist Party cell in a Sheffield steel mill when he leaves the estate.[54] Written in 1926, this version shows that 'Momentarily, Lawrence returned to the working class as a source of alternative values to the harsh and conflict-ridden world of the individual entrepreneur, but it was an uncertain and short-lived optimism.'[55]

To conclude, therefore, it would appear that the geographer is on firm ground when he uses Lawrence's coalfield novels in order to gain a better understanding of place, but we must beware when seeking to use them to gain a better understanding of society. Even with regard to place, however, it should be remembered that, if we accept Parkin's value systems, outlined at the beginning, that 'the subordinate value system restricts man's consciousness to the immediacy of a localised setting', then we must beware also lest we use Lawrence's writings to support accommodative responses to the facts of inequality.

Notes

1. For examples of humanistic and radical geography see D. Ley and M. S. Samuels (eds.), *Humanistic Geography: Prospects and Problems* (Croom Helm, London, 1978); and R. Peet (ed.), *Radical Geography* (Maaroufa Press, Chicago, 1977).

2. R. Peet, 'The Development of Radical Geography in the United States', *Progress in Human Geography*, vol. 1, no. 3 (1977), p. 259.

3. F. Parkin, *Class Inequality and Political Order* (Paladin, St Albans, 1972), pp. 81–2.

4. Ibid., p. 97.

5. Counter-consciousness is a particularly important concept for former colonies which have had their own consciousness suppressed by the colonisers. See, for example, R. Constantino, *Neocolonial Identity and Counter Consciousness* (Merlin Press, London, 1978).

6. R. Hudson, 'Space, Place, and Placelessness: Some Questions Concerning Methodology', *Progress in Human Geography*, vol. 3, no. 1 (1979), pp. 169–74.

7. J. Galtung and M. H. Ruge, 'The Structure of Foreign News' in J. Tunstall (ed.), *Media Sociology* (Constable, London, 1970), pp. 259–98.

8. For a model of perception see, for example, D. C. D. Pocock, 'Environmental Perception: Process and Product', *Tijdschrift voor Economische en Sociale Geografie*, vol. 64, no. 4, (1973), pp. 251–7.

9. Lawrence believed strongly in the need for writing and rewriting. See F. Kermode, *Lawrence* (Fontana, London, 1973), p. 15.

10. For a review of these see Pocock, 'Environmental Perception: Process and Product'; R. M. Downs, 'Geographic Space Perception: Past Approaches and Future Prospects', *Progress in Geography*, vol. 2 (1970), pp. 65–108; M. Pacione, 'Information and Morphology in Cognitive Maps', *Transactions of the Institute of British Geographers*, New Series, vol. 3, no. 4 (1978), pp. 548–68.

11. A brief survey of the Eastwood area is provided by C. M. Sinzelle, *The Geographical Background of the Early Works of D. H. Lawrence* (Didier, Paris, 1964), pp. 8–21.

12. D. H. Lawrence, 'Nottingham and the Mining Country', 1929, in D. H. Lawrence, *Selected Essays* (Penguin, Harmondsworth, 1950), pp. 114–22.

13. Sinzelle, *The Geographical Background*, p. 22.

14. Among the colliers he was, however, a 'butty' with higher status. Ibid., pp. 95–6.

15. H. T. Moore and W. Roberts, *D. H. Lawrence and his World* (Thames & Hudson, London, 1966), p. 11.

16. D. H. Lawrence, *Sons and Lovers*, 1913 (Penguin, Harmondsworth, 1948). There were in fact five children in the family, not four as in the novel. D. H. (Bert, as he was known) was the third son.

17. D. H. Lawrence, *Studies in Classic American Literature* (Heinemann, London, 1924), p. 6.

18. She was the Miriam of *Sons and Lovers*.

19. He taught colliers' sons at the British School in Eastwood in 1902, before becoming a pupil-teacher in nearby Ilkeston from 1903–5. His teaching experiences are those of Ursula in *The Rainbow*. D. H. Lawrence, *The Rainbow*, 1915 (Penguin, Harmondsworth, 1948).

20. T. S. Eliot, in the Foreword to Father W. Triverton, *D. H. Lawrence and Human Existence* (Rockliff, London, 1951), p. viii.

21. F. R. Leavis, *D. H. Lawrence, Novelist* (Chatto & Windus, London, 1955), Appendix; Kermode, *Lawrence*, Prologue; F. B. Pinion, *A D. H. Lawrence Companion* (Macmillan, London, 1978), pp. 16–17.

22. Moore and Roberts, *D. H. Lawrence and his World*, p. 33.

23. Novels such as *Aaron's Rod*, *Kangaroo* or *The Plumed Serpent*, for example, published by Heinemann and Penguin.

24. P. Bentley, *The English Regional Novel* (Allen & Unwin, London, 1941).

25. Quoted in Pinion, *A D. H. Lawrence Companion*, p. 46.

26. S. Sanders, *D. H. Lawrence: The World of the Major Novels* (Vision, London, 1973), pp. 48–53.

27. Ibid., p. 31.

28. Apart from those already cited, others of repute include R. Aldington, *Portrait of a Genius, But . . .* (Heinemann, London, 1950); G. Hough, *The Dark Sun* (Duckworth, London, 1956); H. T. Moore, *The Intelligent Heart* (Heinemann, London, 1955).

29. Quoted in B. Pugh, *The Country of My Heart* (Nottinghamshire Local History Council, 1972), p. 17.

30. Lawrence, *Sons and Lovers*, p. 278.

31. D. H. Lawrence, *The White Peacock*, 1911 (Penguin, Harmondsworth, 1950), p. 153. See L. Spolton, 'The Spirit of Place: D. H. Lawrence and the East Midlands', *The East Midland Geographer, Special Issue in Honour of K. C. Edwards* (1970), pp. 88–96. Spolton notes Gadjusek's count of 145 flowers, plants and trees mentioned in this novel.

32. Sinzelle, *The Geographical Background*, p. 77.

33. Lawrence, *The White Peacock*, pp. 13–15.

34. D. H. Lawrence, *Women in Love*, 1921 (Penguin, Harmondsworth, 1960).

35. Lawrence, *The Rainbow*, pp. 7–8.

36. Lawrence, *Sons and Lovers*, p. 389.

37. Ibid., p. 8.

38. Lawrence, *The Rainbow*, p. 345.

39. D. H. Lawrence, *Lady Chatterley's Lover*, 1928 (Penguin, Harmondsworth, 2nd edn. 1961), pp. 158–9.

40. D. H. Lawrence, *The Lost Girl*, 1920 (Penguin, Harmondsworth, 1950), p. 11.

41. Ibid.

42. Ibid., pp. 65–6.

43. Lawrence, *Women in Love*, pp. 256–7.

44. I. G. Cook, 'Selected Aspects of Environmental Perception in the East Midlands,' unpublished PhD thesis, University of Nottingham, 1977.

45. R. Aldington (ed.), *The Spirit of Place: An Anthology Compiled from the Prose of D. H. Lawrence* (Heinemann, London, 1935), p. vi.

46. Quoted in Sanders, *D. H. Lawrence: The World of the Major Novels*, p. 40.

47. Ibid., p. 69.

48. Ibid., p. 52.

49. Kermode, *Lawrence*, p. 9.

50. Ibid., p. 29.

51. Ibid.

52. Quoted in Pinion, *A D. H. Lawrence Companion*, p. 36.

53. D. H. Lawrence, *The First Lady Chatterley*, 1944 (Penguin, Harmondsworth, 1973).

54. A. Mellor *et al.*, 'Writers and the General Strike' in M. Morris (ed.), *The General Strike* (Penguin, Harmondsworth, 1976), pp. 338–59.

55. Ibid., pp. 347–8.

5 NEWCOMERS, EXISTENTIAL OUTSIDERS AND INSIDERS: THEIR PORTRAYAL IN TWO BOOKS BY DORIS LESSING[1]

David Seamon

An achievement of imaginative literature, writes the phenomenologist Maurice Natanson, is the revelation of 'the experiential foundation of our world'.[2] In this sense, novels, short stories and poems are starting points for phenomenological investigation: they provide descriptive accounts of ordinary and extraordinary human experience which the phenomenologist can explore and order. The aim is an explicit picture of implicit experiential and behavioural patterns delineating the essence of what it means to be and live as a human being. Imaginative literature can be trusted to provide at least the beginnings of such a picture because the microcosm of novel or poem reflects in some measure 'the big world of real life'.[3]

If imaginative literature offers a base from which to generate new phenomenological insights, it also provides a complementary function: a testing ground in which to confirm and amplify existing phenomenological claims. If, as Merleau-Ponty argues, the body works as an intelligent but prereflective consciousness, then accounts in imaginative literature should provide concrete validation of his claim and at the same time offer new evidence and examples.[4] If, as some phenomenologists suggest, at-homeness is an essential aspect of human existence, then this theme should be central in much of imaginative literature, which in turn should provide pictures of at-homeness for the particular place and lifestyle on which the writer has chosen to focus.[5]

This chapter uses two books by the British—African novelist and short-story writer Doris Lessing to explore and extend the phenomenological notions of *existential insideness* and *existential outsideness* developed by the geographer Edward Relph in *Place and Placelessness*.[6] His book investigates the nature of place as an experience and concludes that its essential experiential core is *insideness* — the degree to which a person belongs to and associates himself with place. Relph terms the most profound sense of place *existential insideness*, a situation 'in which a place is experienced without deliberate and selfconscious reflection yet is full with significances'.[7] The experiential opposite of insideness is outsideness, an experience in which the person feels separate from place. Its extreme form Relph calls *existential outsideness*, a situation in which the person feels alienated from an environment devoid of meaning.

Doris Lessing's two books, *In Pursuit of the English* (1960) and *The Four-Gated City* (1969), describe newcomers arriving in a foreign environment and clarify and elaborate Relph's depictions of existential outsideness and insideness.[8] *Pursuit* is a journalistic account of Lessing's first year in London. It describes her departure from Africa and her year of living in a London working-class neighbourhood still bearing the bomb scars of the Second World War. In some ways, *City*, the last novel in Lessing's five-volume series *Children of Violence*, echoes *Pursuit*, since it also describes a young white African woman named Martha Quest coming to live in London after the war.[9] Much more complex and industrious than *Pursuit*, however, *City* explores the nature of human consciousness; it progresses into the present and prophesies the future. I shall use here only the novel's opening chapters, which describe Martha Quest's first few months in London.

The Newcomer and Existential Outsideness

Existential outsideness involves a feeling of homelessness and not belonging. It is the most extreme experience of separation from place. Lessing's experiences in *Pursuit* indicate that the newcomer is an existential outsider who wishes to become an insider in place. Acceptance and at-homeness come only when the newcomer stops converting the place into what he expects it to be and lets it be what it is.

Lessing is thirty when she arrives in London, She is a struggling writer with a three-year-old son. She has come to England because she wants to. 'I can't remember a time', she explains, 'when I didn't want to come to England.'[10] On arrival, however, she finds the place a disappointment and an eyesore; she wishes to leave:

> The White Cliffs of Dover depressed me. They were too small. The Isle of Dogs discouraged me. The Thames looked dirty. I had better confess that for the whole of the first year, London seemed to me to be a city of such appalling ugliness that I wanted only to leave.[11]

A feeling of unreality, Relph claims, characterises the existential outsider, and it is the difference between her image of England and England as she actually finds it that provokes in part this sense of

unreality for Lessing. 'England was for me a grail,' she declares early in
Pursuit, yet she recognises that by its very nature a grail is 'a
quintessence, and by definition, unattainable'.[12] As she travels from
her home in Southern Rhodesia to South Africa, she imagines she will
arrive in England when she gets to Cape Town: 'This is because the Cape
is English, or as the phrase goes, it is pervaded by the remnants of the
old English liberal spirit.'[13] This illusion is shattered as she spends six
weeks in the city, waiting for her ship and living in a dilapidated
boarding house occupied by young English women, all brides of South
African soldiers and waiting for a place to live. The women talk with
bitter homesickness about England and Lessing is 'bored to death'.[14]
She realises she will not encounter England until she arrives there.
'England,' she says, 'I knew by now would not actually begin until the
moment I set foot on its golden soil.'[15]

When she arrives in London, however, the 'golden soil' loses its lustre.
She feels a sense of environmental oppression, which can be seen as
another characteristic of the newcomer's existential outsideness. The
London landscapes and buildings project an atmosphere of isolation,
threat and frustration. Already she has called the White Cliffs of Dover
depressing and London appallingly ugly. The bleakness continues as she
searches for a place to live. While looking for a flat, she walks through
'interminable streets of tall grey houses' that are 'half effaced with
fog'.[16] 'Pale faces' peer up from cellars, 'past rubbish cans', 'railings too
grimy to touch', and 'dirty flights of stairs'.[17]

In the same negative tone, she describes the first house where she
lived for a short time. When she returns to look at the building several
years later it seems 'quite unremarkable', yet when she lived there she
described it in a journal entry as:

> Decaying, unpainted, enormous, ponderous, graceless. When I stand
> and look up, the sheer weight of the building oppresses me. The door
> looks as if it could never be opened. The hall is painted a dead
> uniform cream, that looks damp . . . The flat has six rooms, all
> painted this heavy darkening cream, all large, with high ceilings, no
> sound anywhere, the walls are so thick. I feel suffocated. Out of the
> back windows, a vista of wet dark roofs and dingy chimneys. The
> sky is pale and cold and unfriendly.[18]

If coldness, unkindness and ugliness are part of the newcomer's
existential outsideness, another dimension of unreality of place may be
a sense of confusion and disorder. For Lessing as newcomer, London is

a constant puzzle which she must continuously figure out. 'My head',
she writes, 'was as usual in those early days in London a maze . . . It
seemed to me impossible that the people walking past the decent little
shops that were so alike . . . could ever know one part of London from
another.'[19] As she searches for the address of a flat, she finds the street
she wants is not in her guidebook. Passers-by direct her back and forth,
each saying, 'It's just around the corner.' 'Which corner do they mean?'
she asks herself in bewilderment:

> This business of the next corner is confusing to aliens, who will
> interpret it as the next intersection of the street. But to the
> Londoner with his highly subjective attitude to geography, the
> 'corner' will mean, perhaps, a famous pub, or an old street whose
> importance dwarfs all the intervening streets out of existence, or
> perhaps the turning he takes every morning on his way to work.[20]

In time, suggests Lessing, if the newcomer is unable to make a place for
himself in the new environment, he develops a sense of distrust for
people and place, which gradually may grow and strengthen. How, asks
the newcomer after a period of disappointment, could anything good
come my way in this foreign place? Lessing describes an Australian lady
and her daughter who for several months had searched for a flat large
enough to house the grand piano they had brought with them. They
become so bitter about not finding a place that several times upon
setting out for a possible address, they exclaim, 'What's the use, they
won't have us!' and turn aside into a café to brood over a cup of tea.[21]

The London teashops, says Lessing, are the newcomer's refuge. 'At
a time', she explains, 'when getting a place to live was essential before
we could start to live at all, we would spend the larger part of each
working day . . . sitting in teashops gripped by bitter lethargy.'[22] The
newcomers speak of their home places, unfriendly landlords or
exorbitant London rents. They cannot accept the discrepancy between
what they had hoped for and what they have found. 'We could not
face', says Lessing, 'seeing our fantasies about what we had hoped to
find diminished to what we knew we would have to take.'[23]

If the separation from place that characterises existential outsideness
can foster a sense of alienation and threat for some newcomers, it can
generate a sense of freedom for others. Lessing does not mention this
freedom in *Pursuit*, largely because of her son for whom she must
quickly find a home and means of support. Martha Quest in *City*,
however, is single and unattached when she arrives in London. She has

come partially to find out who she is and, at least for a time, to be different from the Martha Quest of the past. When she meets a stranger and he asks her name, Martha answers, 'Phyllis Jones', and 'for an afternoon and an evening she had been Phyllis Jones, with an imaginary history of wartime work in Bristol'.[24]

Without links to place, especially in a large city like London, the newcomer has a chance to be who and what she wishes to be. Thus Lessing writes about Martha:

> For a few weeks she had been anonymous, unnoticed — free. Coming to a big city for those who have never known one means first of all, before anything else, and the more surprising if one has not expected it, that freedom: all the pressures off, no one cares, no need for the mask. For weeks, then, without boundaries, without definition, like a balloon drifting and bobbing, nothing had been expected of her.[25]

The separation from place, then, that the newcomer feels is potentially both negative and positive. One can be severed from people and place, even threatened and misused. On the other hand, the newcomer is free to experiment with living; she is removed from her past personal history by a place where no one yet knows her. In his discussion of existential outsideness, Relph emphasises its alienating, confusing dimensions because he focuses on the experience of nineteenth- and twentieth-century writers such as Rilke, Proust and Henry Miller, who are perpetually adrift in a world of places 'that are without sense, mere chimeras, and at worst . . . voids'.[26] Relph's probe is only preliminary, however, and by exploring the newcomer's experience one extends the depiction of existential outsideness and details its more positive dimensions.

Existential Insideness

One next must ask how the newcomer becomes an insider and what this means in terms of existential insideness. To become an insider means to reduce one's isolation from place by developing a constellation of experiential ties: a knowledge of how to orient, a feeling for the hidden dimensions of particular places, an understanding of people and events, a sense of personal and interpersonal history in relation to place.

A prime catalyst in this process is the person already at home in

place, the existential insider. Existential insideness is the most intimate
experience of place, signifying the reason why place can be an essential,
even inescapable, dimension of human living and experience.
Existential insideness is a total, unself-conscious immersion in place.
Person and environment are intimately joined by bonds of familiarity,
attachment and at-homeness. Person 'is part of place and it is part of
him'.[27]

Pursuit and *City* provide detailed portraits of existential insiders and
their ways of helping the newcomer feel at home in place. One such
existential insider is Rose, a clerk in a watch shop who lives in the room
next to Lessing's and becomes her first friend in London. Rose's world
is small, geographically. It is 'the half mile of streets where she had been
born and brought up, populated by people she knew; the house where
she now lived, surrounded by *them* – mostly hostile people; and the
West End'.[28] The latter is not a geographical area but a 'fixed journey'
which involves a particular bus route, a certain corner-house, a half-
dozen cinemas, and a walk up Regent Street window-shopping.[29] The
world of Flo, Rose and Lessing's landlady, is even smaller geographically.
It is the basement flat where she and her family live, the food shops
where she is registered for rationing, and the cinema five minutes' walk
away. Flo, says Lessing, 'had never been inside a picture gallery, a
theatre or a concert hall'.[30]

Generally, the existential insider has little interest in the world
beyond his place. Flo speaks of taking an afternoon trip to the Thames,
which she has not seen since before the war. She never goes, however,
at least during the year Lessing lives in her house. Rose has never been
on the other side of the river and has little interest in going. Once
Lessing invites her to come on a bus trip there. Rose considers the idea
for a week but refuses: 'I don't think I'd like those parts,' she says, 'not
really. But you go and tell me about it after.'[31]

Though it has a definite geographical content, the world of the
existential insider is not so much an explicit place as it is an unself-
conscious, taken-for-granted part of daily living inseparable from the
rest of the insider's day-to-day world. 'Have you always lived in
London?' Lessing asks Rose when she first meets her. There is a pause
before Rose answers. 'It was', realises Lessing, 'because she found it
difficult to adapt herself to the idea of London as a place on the map
and not a setting for her life.'[32] There is a small, grudging note in Rose's
voice as she replies that she has always lived in London. This tone,
Lessing realises, is 'the most delicate of snubs, as if she were saying:
It's all very well for you. . .'[33] The outsider, automatically interpreting

place as an objective entity, collides with the insider's unquestioned, taken-for-granted attitude toward place.

In time, Lessing comes to understand that the world Rose has created in the impersonality of London is 'a sort of tunnel, shored against danger by habit, known buildings, and trusted people'.[34] Habits revolve around recurrent behaviours and routines. As Lessing and Rose walk through the area where Rose grew up, Lessing becomes bewildered and disoriented, but Rose moves 'along the street without seeing it, her feet quick and practised on the pavement',[35] an insider moving easily and fluidly in her place.

There is also a regularity of events: for example, eating out at the same corner-house once a month, going down to Flo's basement for Sunday dinner each week, or helping her clean up each night. Habit underlies these events. 'Do you have to help her?' asks Lessing in regard to Rose's nightly dishwashing for Flo. 'No, not really,' answers Rose. 'But I've gotten into the habit of it . . . you must watch yourself and don't let yourself get into the habit of doing things. I'm telling you for your own good.'[36]

A large portion of day-to-day living for the existential insider is grounded in such habitual routines as washing-up or walking the same way between home and work each day. Relph suggests that such patterns may lead to a 'sheer drudgery of place, a sense of being tied inexorably to *this* place, of being bound by the established scenes and symbols and routines'.[37] For the outsider, especially, the requirements of habit can seem monotonous and stultifying, but often the insider accepts them without question and insists they continue in the same way each time. Lessing describes Rose's taking-a-night-out routine:

> We always walked to the same bus-stop, and it had to be the same bus-stop, and the same bus, though there were several which would have done. She kept pulling me back, saying: 'No, not that bus. That's not the number I like.' And if the bus did not have seats free, downstairs, on the left-hand side, she would wait until one came that had. She made me sit near the window. She liked to sit on the aisle. . .[38]

At the corner-house to which they go, there is always a queue, which Lessing as an outsider finds tedious. For Rose, however, the queuing is unnoticed and unquestioned — a necessary part of the 'night out'. Once inside, Rose makes sure they sit by the band and always order the same fare: beans on toast, with chips and bacon. Lessing once asks her why; Rose replies that it reminds her of the canteen food during the war.[39]

Habit, routine and ritual lay the groundstones in the existential insider's world for a matter-of-fact knowledge of people and places, to which extend emotional bonds. As Lessing and Rose walk through her childhood neighbourhood, Lessing realises that 'this part of the great city was home, to her; a different country from the street, not fifteen minutes' walk away, where she now lived'.[40] Being in this place 'reinstated her as a human being with rights of possession in the world'.[41] Rose is 'happily nostalgic'; as she and Lessing turn into the street where she grew up, she looks 'lovingly around her'.[42] Each place in Rose's home area has a special meaning for her: passing a shop, she says, 'I used to get all my shoes there.' Or: 'Before the war they sold a bit of fried skate in this shop better than anything.'[43] In the same way, Rose is intimately familiar with the people of her place: 'She knew every face we saw in the area we lived in, and if she did not, made it her business to find out.'[44]

Walking through Rose's home neighbourhood, Lessing feels she is under her protection: 'She kept herself between me and the crowd and at every moment she nodded and smiled to some man or woman leaning against a counter or stall.'[45] At a fruit stand, Rose buys cherries and hands the seller exact change. He returns a threepence. 'Thanks,' says Rose. 'And this is my friend, see? She'll be coming down here, I expect so you treat her right.'[46] As they leave Rose tells Lessing to buy her fruit there: 'Now he knows you're my friend, it'll be different. And don't you go buying stuff from those barrows. That's only for those who don't know better. I mean you have to know which barrows are honest.'[47]

Penetrating beneath the more readily identifiable familiarity of routines, places and people is a less tangible, more difficult to describe dimension of existential insideness: an invisible, generally unspoken matrix of feelings and vibrations whose essence is hinted at by such phrases as 'sense' or 'spirit' of place. Martha Quest, early in her London experience, lives for several weeks with an elderly couple, Jimmy and Iris, who run a café. At first this place for Martha is nothing more than one of the millions of little shops, 'each one the ground floor of an old house', all over London.[48] In time, however, Martha becomes sensitive to the atmosphere of this place and its people:

> It was jolly in the cafe; people coming in knew each other, knew Iris and Jimmy. They had shared, many of them, their childhood, their

lives. They had shared, most of them, the war. And they had opened their hearts to her.[49]

'What people actually said in that cafe', notes Lessing, 'was the least of what they were able to convey.'[50] As Martha sits there one morning, preparing to leave Jimmy and Iris and live elsewhere, she has grown sensitive enough to this place to identify unspoken meanings and identify the café's mood: the air 'vibrates with interest, tact, sympathy — friendship, in short; all the pressures for a blissful few weeks since Martha had been in England, rather London, she had been freed from'.[51] The outsider moves toward insideness when he can 'tune in' to the unseen vibrations and nuances of place, which perhaps more than anything else are the crucial cement of existential insideness.

The invisibility of place may also possess temporal dimensions. The café has a special rhythm which repeats day after day. At five, Iris opens the café unofficially and feeds 'cornflakes, toast, scrambled egg and tea to some lorry drivers from a lodging house down the street whose landlady would not feed them so early'.[52] There are also a few older men and charwomen there. At nine, 'the side of the card that said OPEN was turned in invitation to the pavement'.[53] Customers are officially received. In her weeks working at the café, Martha becomes aware of its rhythm: between five and eight, for example, it is 'a scene of bustling steaming animation, of intimacy'.[54] As she approaches the café for the last time one morning around eight, Martha notes 'against the dim muslin that screened the cafe, shapes of bodies; the lively intimacy of the early-morning session shed warmth onto the pavement'.[55] A part of Martha regrets that she must leave.

Insiders and Outsiders

For the existential insider, place is more than its material parts. It is not an objective entity or an environment to be explored and used as it may be for the outsider. Rather, place is a setting of invisible, shifting energies which the insider understands without thinking about it. Existential insiders understand their place through continuous, day-to-day living in that place. They can initiate outsiders like Lessing and Martha to the hidden meanings of place and help them feel inside — at least more so than they were before the insider's help. Assemble the experiences of existential insiders, says Lessing, and there would be a recording instrument, a sort of six-dimensional map which included the histories and lives and loves of people, London — a section map in depth. This is where London exists, in the minds of people who have lived in such and such street since they were born. . .[56]

Perhaps the best example is Iris, whose knowledge of her neighbour-hood is a profound familiarity and concern for the simple, commonplace parts of place which for Martha would otherwise go unnoticed or seem inconsequential. When she walks with Iris through the bomb-torn London streets that are her home, Martha sees

in a double vision, as if she were two people: herself and Iris, one eye stating, denying, warding off the total hideousness of the whole area, the other, with Iris, knowing it in love. With Iris, one moved here, in a state of love, if love is the delicate but total acknowledgement of what is.[57]

Love . . . the delicate but total acknowledgement of what is. Herein lies the essential difference between newcomer and existential insider. The newcomer cannot acknowledge place because it appears only as a surface; he cannot identify or differentiate its parts nor contact the hidden nuances that permeate place and make it for Iris and other existential insiders a kind of experiential synergy vibrating with implicit personal and interpersonal meaning. 'A whole,' writes Lessing. 'People in any sort of communion, link, connection, make up a whole.'[58]

For Martha, Iris's neighbourhood is only what it appears: a bomb-scarred slum offending the senses. In contrast, Iris's encounter with place penetrates beneath the realm of simple seeing and touches a storehouse of feelings, remembrances and events all preserved in a fabric of fondness and concern:

Passing a patch of bared wall where the bricks showed a crumbling smear of mushroom colour, Iris was able to say: Mrs. Black painted this wall in 1938, it was ever such a nice pink. Or, looking up at a lit window, the curtains drawn across under the black smear of the blackout material which someone had not got around to taking down: Molly Smith bought those curtains down at the market the first year of the war, before things got so scarce. Or, walking around a block in the pavement, she muttered that the workmen never seemed to be able to get that piece in square, she always stubbed her foot against it.[59]

Why does the insider involve himself with the outsider? In part it is coincidence: outsider accidentally meets insider who offers assistance. Lessing coincidentally meets Rose in the watch shop because her watch breaks; Martha meets Iris and Jimmy 'by chance flopping down in the cafe for a cup of tea, her legs having collapsed from hours of walking'.[60] In addition, the insider may help the outsider partially out of a vague sense of the outsider's unfamiliarity with place and people. 'Of course, you're a foreigner, and don't know yet,' says Rose as she gives Lessing advice on renting a room from Flo.[61] Further, the insider may be drawn to the sense of difference the outsider projects: 'The reason I like you,' explains Rose to Lessing, 'well, apart from being friends now, it's because you say things that make me think.'[62] Another example is Stella, a woman living in the dock area of London's South Bank. Accidentally spotting the foreign-looking Martha 'in green linen, sandals and sunburn' on the wharves one day, Stella asks her to tea, and Martha ends up living there a month.[63] Stella takes her in 'because of an unfed longing for travel and experience which was titillated every minute by the river, by the ships that swung past her windows, by the talk of foreign countries'.[64] Stella wishes Martha, the outsider and stranger, to talk about foreignness,

> and Martha, feeling that nothing in her experience could match up to such an appetite for the marvellous, made a discovery: that it was enough to say, the sun shines so, the moon does thus, people get up at such an hour, eat so and so, believe such and such – and it was enough. Because it was different.[65]

A working relationship between insider and outsider does not always come easily. Often there is confusion and misinterpretation. When, for example, Lessing has known Rose for only a few days, she goes out for an evening walk alone, not thinking Rose might be hurt because she is not invited. The outsider assumes that insiders are comfortably settled in place; to ask them to partake in the outsider's life seems an intrusion: 'I did not know she wanted to come with me. Coming to a new country, you don't think of people being lonely, but having full lives in which you intrude.'[66]

For the insider, the newcomer sometimes seems naive and stupid, behaving and acting in an improper manner yet largely unaware of his profound ignorance of place. For example, several months after Rose and Lessing have become friends, Rose explains the difficulty she had in accepting her and understanding who she is:

When you first came to live with us . . . you just made me sick. It
wasn't that you fancied yourself, it wasn't that, but you was just
plain ignorant about everything. You didn't know nothing about
anything, and you didn't even know you were ignorant. You made
me laugh, you did really.[67]

Sometimes, it is not an insider but an outsider turned insider who
helps the outsider understand place. This person provides an introduction
to place founded not in immersion and invisibility but rather in empathy
and a growing appreciation of place. In *Pursuit*, this person is Miss
Privet, a well-educated lady from the Midlands who comes to live in a
room upstairs from Lessing. Miss Privet, a prostitute, reads Pepys'
Diary and often walks over the streets he walked. 'Nothing's changed
much has it?' she says to Lessing.[68] Explaining that she has still not
learned to like London (this said almost a year after she has arrived),
Lessing is told by Miss Privet that it takes time to accept and appreciate
a place. She offers to show Lessing around London and then rushes
upstairs to fetch a print of Monet's *Charing Cross Bridge*. 'That's
London,' she says, 'but you have to learn to look.'[69]
The next evening Lessing is taken by Miss Privet to her favourite
place in London – the view of Trafalgar Square from the steps of the
National Gallery. Lessing sees London anew:

It was a wet evening, with a soft glistening light falling through a low
golden sky. Dusk was gathering along walls, behind pillars and
balustrades. The starlings squealed overhead. The buildings along
Pall Mall seemed to float, reflecting soft blues and greens on to a wet
and shining pavement. The fat buses, their scarlet softened, their
hardness dissolved in mist, came rolling gently along beneath us,
disembarking a race of creatures clad in light, with burnished hair
and glittering clothes. It was a city of light I stood in, a city of bright
phantoms. But Miss Privet was not one to harbour her pleasures
beyond reasonable expectation. For ten minutes I was allowed to
stand there while the light changed and the thin clouds overhead
sifted a soft, drenching golden atmosphere.
Then she said, 'Now we should go. It'll be dead in a minute, just
streets.'[70]

Implications

However intimate the newcomer becomes with place, he can never
become a complete insider because his past permeates and colours the

present place. All foreigners, says Lessing, no matter how long in their adopted place, are susceptible to the feeling 'that we shall ever be aliens in an alien land'.[71] In today's modern world, the rate of technological change and ease of geographical mobility make this fact applicable to more and more people. The result is an erosion of existential insideness and places, both of which require a rootedness and stability for their continued existence. More and more people feel a sense of outsideness, often alienation; environments are objects severed from person or landscapes of placelessness.[72]

One practical value of exploring insideness and outsideness, therefore, is that it sensitises the student to these experiences and helps him to better understand and accept them in his own life – as he leaves his parents' home for university, visits a foreign country, or changes jobs and moves to a new place. Second, such understanding attunes the outsider who would like to promote change in a particular place to the inertia of life in that place. As Rose says to Lessing: 'I've been thinking about you, dear. Your trouble is this. You think all you've got to do is say something, and then things'll be right. Well, they won't be.'[73]

Devising and formulating plans for places does not necessarily lead to change, especially if that change is more than the existential insider can bear. Immersed in place, he often resists change or adjusts badly. The outsider must do more than think and talk about change: he must find a way to inform the insider gently that his place might be other than it is. Contrasting his position with the world of insideness he wishes to help may lead the outsider to a more viable course of action, particularly if he can participate in the insider's world, understanding his needs and point of view directly as Lessing and Martha Quest do.

In more general terms, the phenomenological study of imaginative literature provides new approaches and notions for research in environmental behaviour and experience. The imaginative writer, says Tuan, 'is forced to think through in detail the effects of events and initiatives on the densely textured world he has created'.[74] Out of such a precise and demanding exercise, the social scientist 'can learn to ask questions and formulate hypotheses'.[75] In this sense, the notions of insideness and outsideness offer a valuable new perspective in the study of environmental behaviour and experience. Imaginative literature can help to extend and clarify this perspective, grounding it in the lives of particular people and places described with the lucidity and precision of good imaginative writing.

The world today faces the problem of 'future shock' on the one hand, the questions of energy and ecology on the other. The roots of both of these dilemmas are to be found in people's experiential relationship with place and environment. Perhaps the major need is the ability to be both insider and outsider: to feel at home in a particular place, yet to understand that place as it is part of a larger earth whole. Exploring insideness and outsideness, especially as they are described in imaginative literature, may help the student to be citizen of both world and particular place.

Notes

1. This chapter is a revised version of a presentation by the same title given at the annual meeting of the Institute of British Geographers, Manchester, 4 January 1979, for the special session 'Geography and Literature', organised by D. C. D. Pocock.

2. M. Natanson, *Literature, Philosophy and the Social Sciences* (Martinus Nijhoff, The Hague, 1962), p. 97.

3. Ibid., p. 88. Discussions of the value of imaginative literature to geography include D. Seamon, 'The Phenomenological Investigation of Imaginative Literature' in G. T. Moore and R. G. Golledge (eds.), *Environmental Knowing: Theories, Research and Methods* (Dowden, Hutchinson & Ross, Stroudsburg, Pa., 1976), pp. 286–90; Yi-Fu Tuan, 'Literature, Experience and Environmental Knowing' in Moore and Golledge, pp. 260–72; Yi-Fu Tuan, 'Literature and Geography: Implications for Geographical Research' in D. Ley and M. S. Samuels (eds.), *Humanistic Geography: Prospects and Problems* (Croom Helm, London, 1978), pp. 194–206; C. L. Salter and W. J. Lloyd, 'Landscape in Literature', *Resource Papers for College Geography* (Association of American Geographers, Washington DC), no. 76–3 (1977); D. C. D. Pocock, 'The Novelist's Image of the North', *Transactions of the Institute of British Geographers*, vol. 4 (1979), pp. 62–76. On the methods of phenomenology and an empirical demonstration of its value to geography, see David Seamon, *A Geography of the Lifeworld* (Croom Helm, London, 1979).

4. M. Merleau-Ponty, *The Phenomenology of Perception*, translated by Colin Smith (Humanities Press, New York, 1962).

5. For example, Frank Buckley, 'An Approach to a Phenomenology of At-Homeness' in A. Giorgi, W. Fischer and R. von Eckartsberg (eds.), *Duquesne Studies in Phenomenological Psychology*, vol. 1 (Duquesne University Press, Pittsburgh, 1971), pp. 198–211; Anne Buttimer, 'Home, Reach, and a Sense of Place' in Anne Buttimer and David Seamon (eds.), *Place and Journey: Excursions in Human Geography* (Croom Helm, London, 1980); Seamon, *Lifeworld*, note 3.

6. E. C. Relph, *Place and Placelessness* (Pion, London, 1976).

7. Ibid., p. 55.

8. Doris Lessing, *In Pursuit of the English* (Popular Library, New York, 1960); Doris Lessing, *The Four-Gated City* (Knopf, New York, 1969). Quotations in this chapter from the latter book are drawn from the Bantam edition, New York, 1970.

9. The other volumes in the series are *Martha Quest* (1952), *A Proper Marriage* (1954), *A Ripple from the Storm* (1958) and *Landlocked* (1965). Paperback editions of these books are published by MacGibbon and Key, London.

10. Lessing, *Pursuit*, p. 12.

11. Ibid.

12. Ibid., pp. 13 and 11.

13. Ibid., p. 15.
14. Ibid., p. 25.
15. Ibid.
16. Ibid., p. 40.
17. Ibid.
18. Ibid., pp. 32–3.
19. Ibid., pp. 47–8.
20. Ibid., p. 40.
21. Ibid., p. 39.
22. Ibid.
23. Ibid., p. 40.
24. Lessing, *City*, p. 17.
25. Ibid., p. 4.
26. Relph, *Place and Placelessness*, p. 51.
27. Ibid., p. 55.
28. Lessing, *Pursuit*, p. 101.
29. Ibid.
30. Ibid.
31. Ibid., p. 102.
32. Ibid., p. 54.
33. Ibid.
34. Ibid., p. 101.
35. Ibid., p. 54.
36. Ibid., p. 76.
37. Relph, *Place and Placelessness*, p. 41.
38. Lessing, *Pursuit*, p. 103.
39. Ibid., p. 104.
40. Ibid., p. 74.
41. Ibid.
42. Ibid., p. 75.
43. Ibid., pp. 74–5.
44. Ibid., p. 101.
45. Ibid., p. 74.
46. Ibid.
47. Ibid.
48. Lessing, *City*, p. 8.
49. Ibid., p. 12.
50. Ibid.
51. Ibid., p. 4.
52. Ibid., p. 75.
53. Ibid.
54. Ibid.
55. Ibid., p. 77.
56. Ibid., p. 10.
57. Ibid.
58. Ibid., p. 221.
59. Ibid., p. 10.
60. Ibid., p. 12.
61. Lessing, *Pursuit*, p. 61.
62. Ibid., p. 73.
63. Lessing, *City*, p. 14.
64. Ibid.
65. Ibid., p. 15.
66. Lessing, *Pursuit*, p. 63.
67. Ibid., p. 73.

68. Ibid., p. 229.
69. Ibid.
70. Ibid., pp. 229–30.
71. Ibid., p. 10.
72. Relph, *Place and Placelessness.*
73. Lessing, *Pursuit*, p. 91.
74. Tuan, 'Literature and Geography', p. 201, note 3.
75. Ibid.

6 ROOTS AND ROOTLESSNESS: AN EXPLORATION OF THE CONCEPT IN THE LIFE AND NOVELS OF GEORGE ELIOT

Catherine A. Middleton

A human life, I think, should be well rooted in some spot of a native land, where it may get the love of tender kinship for the face of the earth, for the labours men go forth to, for the sounds and accents that haunt it, for whatever will give that early home a familiar unmistakable difference amidst the future widening of knowledge: a spot where the definiteness of early memories may be inwrought with affection, and kindly acquaintance with all neighbours, even to the dogs and donkeys, may spread not by sentimental effort and reflection, but as a sweet habit of the blood.[1]

This is one of George Eliot's last statements of her feeling that one of the necessary conditions for meaningful human existence is an attachment to a specific place. Attachment to a particular place gives a person security in his uniqueness by enabling him to appreciate some of the sources from which his life springs; roots in a particular place at a particular time influence a person's life by encouraging his growth and development while providing a firm emotional, social and intellectual foundation on which to build.

It is essential to stress the importance of factors other than purely physical ones in this idea of rootedness, for attachment to a particular place is compounded of many different relationships: a person who is 'rooted in some spot of a native land' is rooted not only in a geographical landscape but also in a social landscape (in a community), in an emotional landscape (in a family or in intimate relationships with a few individuals) and in an intellectual landscape (in the knowledge and ideas which he has acquired). An individual is also located in a temporal landscape; his life in a particular place has been lived at a particular time.

An individual is rooted in a place when he has developed these relationships to such an extent that he finds it uncomfortable to move from the place; if he does move, the features of this landscape, compounded of so many different elements, will continue to affect the

101

way he thinks, feels and acts. This place, experienced in the past, will become a reference point against which all his subsequent experiences of place are measured. As Eliot wrote: 'These are the things that make the gamut of joy to Midland-bred souls — the things they toddled among, or perhaps learned by heart standing between their father's knees while he drove leisurely.'[2] This chapter cannot attempt to deal with the large question of why people move away from the places in which they have felt rooted, but I should like to explore the idea of rootedness and rootlessness as it appears in the life and work of George Eliot.

An Introduction to Eliot's Life and Literary Technique

Mary Ann Evans (who later used George Eliot as her *nom de plume*) was born at South Farm, Arbury Lane, Chilvers Coton, Warwickshire, on 22 November 1819. She moved to Griff House on the Coventry Road, where it is joined by Arbury Lane, in 1820. In 1828 she went to school at Nuneaton, returning to Griff House to look after her father when her mother died in 1836. When her brother Isaac married in 1841, Mary Ann and her father moved to Bird Grove, 'set well back from the Foleshill Road, surrounded by fine trees', less than a mile from Coventry. She remained at Foleshill until her father's death in 1849.

After her father's death Mary Ann (now calling herself Marian) spent eight months in Geneva. On returning to England she spent some time moving backwards and forwards between Coventry and London, finally making London her 'base' in 1851. She stayed at 'Rosehill' (the home of her Coventry friends Charles and Caroline Bray) for the last time in June 1854. In that year she began to live with George Henry Lewes.[3] For the remainder of her life (she died in 1880) London was to be her home, although she and Lewes travelled frequently in England and on the Continent.

George Eliot's rootedness in the Midlands is expressed in her novels in different ways. Changes in the way in which she discusses places may be related to her increasing temporal distance from the places in which she spent her childhood. It is equally important, if not more so, to relate Eliot's techniques of writing about places and their importance to individuals to her increasing maturity as a writer concerned, primarily, with the relationships between individuals.

The most noticeable development in George Eliot's technique of writing about places is a general tendency for the amount of topographical detail included to decrease as her writing matures.

Table 6.1: Date of Publication, with Temporal and Geographical Settings, of Novels of George Eliot

Title	Publication date	Duration of narrative	Location of setting
Scenes of Clerical Life			
'Amos Barton'	1858	1830–2	Chilvers Coton
'Mr. Gilfil's Love-Story'	1858	1780s–90s	Chilvers Coton
'Janet's Repentance'	1858	c. 1830	Nuneaton
Adam Bede	1859	1799–1801	N.E. Staffordshire and S.W. Derbyshire
The Mill on the Floss	1860	c. 1830	Lincolnshire (with many elements of the Warwickshire landscape)
Silas Marner	1861	Napoleonic Wars	Central Midlands
Romola	1863	1492–early 1500s	Florence
Felix Holt	1866	1830–2	Warwickshire
Middlemarch	1871–2	1830–2	Coventry and surrounding parishes, and Rome
Daniel Deronda	1874–6	1865–6	Wessex, London and the Continent, especially Leubronn and Genoa

Several causes could be cited to account for this: these include a loss of memory of topographical detail (when she is writing about the Midlands), decreasing interest in topographical detail, or an increasing feeling that such detail does not form an essential fraction in the novel and may even, by distracting the reader's attention away from the main issues with which the author is concerned, be undesirable from a literary point of view. It is not really possible to decide which, if any, of these causes is the most likely explanation for the decrease in topographical detail in Eliot's novels; perhaps each cause operated in some degree.

I should like to make one more point on this subject: in her earliest works Eliot drew very heavily on the details of the landscapes she knew as a child, setting her characters in ready-made environments. As her analysis of character and the relationships between individuals grew more sophisticated, so did her treatment of the geographical setting of her novels. Increasingly, instead of relying on particular places to form a setting for her novels, Eliot derived her own partly fictitious, partly 'real' settings by observing and mingling elements from many different landscapes and, indeed, by 'inventing' topographical detail.

It would also be true to say, I think, that different novels emphasise different aspects of the idea of rootedness. Thus, as has been indicated above, rootedness comprises many elements – including physical, emotional and social factors; and it is possible to see, to some extent, Eliot attempting to justify and make sense out of her own removal from the place in which she spent her childhood by the development of a greater emphasis on the need for roots in a social environment and in relationships with other particular individuals. To a considerable extent one can see *Middlemarch* as an exploration of the idea that emotional rootedness, perfect love and understanding between individuals, can be a substitute for physical and social rootedness. Yet in her last novel, *Daniel Deronda*, Eliot gives us Gwendolen Harleth, who lacks any of these elements of rootedness, and Deronda himself – to whom Eliot seems to extend wholehearted approval – who is desperate to establish a physical location, a place for the growth of roots, for a whole race.

The remainder of this chapter discusses some of these ideas in the context of the individual novels themselves; Table 6.1 provides a general factual introduction to the novels, their dates of publication, and temporal and geographical settings.

Scenes of Clerical Life

The importance of the temporal settings of George Eliot's novels, in terms of rootlessness, has been admirably discussed by Auster, who has written:

> She stands really between two worlds: one is the stable, coherent society of the Midlands, which she remembers from her youth and to which she is powerfully drawn, even as she rejects its parochialism, and the other half is the more or less modern world of the second half of the Nineteenth Century, to which she is committed by her intellectual interests, by almost her whole personality, in fact, but a world whose rootlessness and restlessness, superficiality and discontinuity, make her greatly anxious.[4]

Eliot sees the period of her childhood (despite its Reform Act and major industrial and social changes) as a period of stability; a time when people had close and meaningful relationships with the land on which they lived and with the people to whom they related. In contrast, after moving from her childhood home to live in London and travel abroad (in the 1850s), she finds in the world in general symptoms of a lack of stable living, a restlessness particularly visible in the way in which people appear to be reluctant to remain in one place.

As early as 'Janet's Repentance' Eliot explores this idea; yet here she suggests ironically, in comparing the present with the past, that although Milby now has 'a handsome railway station, where the drowsy London traveller may look out by the brilliant gas-light and see perfectly sober papas and husbands alighting with their leather-bags after transacting their day's business at the county town',[5] these changes do not necessarily imply equivalent developments in terms of moral growth. For although she writes at the beginning of the book that 'Milby is now a refined, moral, and enlightened town, no more resembling the Milby of former days than the huge, long-skirted, drab great-coat that embarrassed the ankles of our grandfathers resembled the light paletot. . .' we are also told that when the town was previously 'dingy-looking'[6] it was one in which moral growth could occur. The work of salvation done by Edgar Tryan is best seen in Janet Dempster 'rescued from self-despair, strengthened with divine hopes, and now looking back on years of purity and helpful labour'.[7] Eliot's rootedness in past places and times is indicated by her emphasis on the idea that technological advance is not necessarily associated with emotional, moral and social advance. Janet's rootedness in her physical environment is hardly commented on: it is her social and environmental roots which are called into question and, eventually, brought into a satisfactory state. In one of the other *Scenes*, however, Eliot shows a more specific interest in the question of physical rootedness.

Apart from the exception noted above, Eliot in her early novels ignores the discrepancy between past and present worlds by ignoring the present and locating the novels firmly in the landscapes of her youth. This precision extends to topographical details. In the *Scenes*, 'Shepparton' is closely similar to the parish of Chilvers Coton, where Eliot was brought up. There is a meticulously accurate (bordering on the boring) description of Arbury Hall, here called 'Cheverel Manor', and 'Milby' is Eliot's Nuneaton. Even here topographical unrealities may be detected. Cheverel Manor is said to be some miles from Shepparton, a detail which conflicts with the actual distance of Arbury Hall from Chilvers Coton. The manor is described as undergoing the same physical changes at the same time as Arbury Hall (although Eliot here accepts this change as not upsetting the essential stability of the times): 'Beginning in 1750 . . . Sir Roger Newdigate . . . rebuilt the huge hollow square of the Tudor house, vaulted the cloister in the inner court, added oriel windows, turrets, pinnacles, and castellated battlements in Gothic style. . .'[8] Sir Christopher Cheverel began his work on Cheverel Manor in 1775, and

> For the next ten years Sir Christopher was occupied with the architectural metamorphosis of his old family mansion; thus anticipating, through the prompting of his individual taste, that general reaction from the insipid imitation of the Palladian style, towards a restoration of the Gothic, which marked the close of the Eighteenth Century.[9]

With this interest in the part played by Cheverel Manor in Sir Christopher's sense of rootedness, Eliot shows a belief in the power of the physical environment to be itself supremely important in giving an individual a sense of attachment, a feeling of belonging. In the case of Amos Barton physical roots and social roots are closely linked. Amos was not greatly popular with his parishioners until the death of his wife 'called out their better sympathies' which resulted in the development of 'a real bond between him and his flock'.[10] The death of the children's mother enabled some of Amos's parishioners to take a more personal interest in his family, thus establishing them in the community. Eliot writes the following when Amos hears that he is to leave Shepparton, where he lived close to Milly's grave: 'To part with that grave seemed like parting with Milly a second time; for Amos was one who clung to all the material links between his mind and his past.'[11] In a tragic way, part of Amos's life has become part of the physical environment in which he is now rooted.

Adam Bede

In *Adam Bede* (1859) Eliot retains her technique of precise physical location; the landscape this time, however, is one more familiar to her through tales told by her father than through her own experience:

> As to my indebtedness to facts of locale . . . connected with Staffordshire and Derbyshire — you may imagine what kind that is, when I tell you that I never remained in either of those counties

more than a few days together, and of only two such visits have I
more than a shadowy, interrupted recollection. The detail which I
knew as facts and have made much use of for my picture were
gathered from such imperfect allusion and narrative as I heard from
my father in his occasional talk about old times.[12]

Yet when the book appeared the village of 'Hayslope' was immediately
recognised as Ellastone in North Staffordshire. According to Auster,

Not only is the description of Loamshire and Stonyshire a
recognisable account of north-eastern Staffordshire and south-
western Derbyshire, but the fictional places like Hayslope,
Oakbourne, and Snowfield closely resemble in location, appearance
and relative situation the real places . . . that were quickly identified
as their prototypes.[13]

Since this accurate source of information was her father, it is fitting
that the novel should be set in the period of her father's young
manhood, a period which Eliot conceives of as being, if anything, even
more stable than the time of her own youth. Eliot emphasises the need
for physical roots by keeping the action almost entirely within the area
of Hayslope; the only major journey on which we are taken is Hetty's
attempt to find Donnithorne during her pregnancy – and this is
important as a symbol of Hetty's alienation from and lack of roots in
her home, family and community. Not content to love Adam, the man
who is deeply rooted in his environment, Hetty is seduced by Arthur
Donnithorne, heir to the Squire. Eliot writes: 'I think she had no feeling
at all towards the old house, and did not like the Jacob's Ladder and the
long row of hollyhocks in the garden better than other flowers –
perhaps not so well.'[14] After pregnancy has forced her to leave home to
travel through strange places, she begins to realise the importance of
having roots:

for the first time, as she lay down . . . in the strange hard bed, she
felt that her home had been a happy one, that her quiet lot at
Hayslope among the things and people she knew . . . was what she
would like to wake up to as a reality.[15]

Hetty's recognition of the need for roots is supported and reinforced by
Mr Poyser. Threatened with eviction from his home, he remarks that
'I should be loathe to leave the old place, and the parish where I was

bred and born, and father afore me. We should leave our roots behind us, I doubt, and niver thrive again.'[16] Dinah feels a sense of responsibility towards the place which has nourished her: 'I'm not free to leave Snowfield, where I was first planted, and have grown deep into it, like the small grass on the hill-top.'[17]

The Mill on the Floss

The point made by Mr Poyser in *Adam Bede* is made again in Eliot's next novel, *The Mill on the Floss* (1860), by Mr Tulliver, also threatened with eviction, from his mill. Eliot writes that Mr Tulliver

> couldn't bear to think of himself living on any other spot than this, where he knew the sound of every gate and door, and felt that the shape and colour of roof and weather-stain and broken hillock was good, because his growing senses had fed on them.[18]

Auster describes this as 'an almost organic attachment to the Mill'.[19]

In this novel Eliot begins, more seriously, to try to sort out the conflicts she finds between the new world and the old. She seems to conclude, on the whole, that the best individuals — the Tullivers — are unable to adapt to the rapid change and restlessness of the new order. Auster suggests that the sense of change that is built up in the course of the novel is primarily centred on economic matters and, to a smaller extent, social ones.[20] Deane, for example, eulogises on trade, the firm, and steam: 'It's the steam, you see, that drives on every wheel double pace, and the world of fortune along with 'em.'[21] On the other hand, Tulliver's ability to cope with life is whittled away; he remarks that 'if the world had been left as God made it, I could ha' seen my way'.[22]

There is some ambiguity, however, in this novel. For while the new order, with its restlessness and rootlessness, is unsatisfactory, the old order is shown to be deficient unless it is transformed by particular qualities of character. Eliot forgets about change altogether while she is describing the effects of the drowning of Tom and Maggie:

> The desolation wrought by that flood, had left little visible trace on the face of the earth, five years after. The fifth autumn was rich in golden cornstacks, rising in thick clusters among the distant hedgerows: the wharves and warehouses on the Floss were busy again, with echoes of eager voices, with hopeful lading and unlading.[23]

While the flood enables Maggie to prove to Tom that 'in her passionate, romantic fitfully visionary idealism she has struggled to absorb and preserve the personal integrity and responsibility that ennoble the otherwise drab moral order' of her family,[24] the social and geographical order of St Oggs is not undermined. Maggie, perhaps, is too much of a visionary idealist to survive in a social environment of ordinary individuals: and instead of leaving Maggie in some state of rootlessness and restlessness because of her moral superiority to the individuals by whom she is surrounded, Eliot allows her to drown.

The strength of Maggie's rootedness in her home is never in question, and many of the places she loves bear close resemblances to the places Eliot loved in her youth; but as Eliot's perception that the equations of old equals good and new equals bad may be oversimplifications of what was really happening, so she allows the distinctions between one topographical detail and another to blur. The novel is set, primarily, in Lincolnshire. Eliot stayed at Gainsborough, which became St Oggs in the novel. Lewes wrote that Eliot was 'wanting to lay the scene of her new novel on the Trent . . . [and] we took a boat and rowed down the Idle, which we ascended on foot some way, and walked back to Gainsborough'.[25] Dorlcote Mill, the home of the Tullivers, was taken from Eliot's recollections of her first home, South Farm.[26] Many of the descriptions are also from Warwickshire; the Round Pond of the novel bears the same name as a pool near Griff House, of which Eliot was fond. According to Haight, 'the flat country along the Trent lacked the picturesque wooded lanes that Maggie haunted, and the Red Deeps with their grand fir trees'.[27] Therefore Eliot transported those aspects of her home environment to the banks of the Trent.

Eliot had not visited her home area since 1854; by 1860 she was, in part deliberately, blurring the details of the remembered landscape of her youth and mixing them with details culled from other places. In some ways Maggie's rootedness in this landscape seems like a memorial to the places — now composing a set of details not entirely conforming to 'reality' — to which Eliot herself felt most attached. While both felt deeply rooted in a particular area, neither Maggie nor Eliot was able to develop emotional, social, moral and intellectual roots in the place: Maggie died and Eliot moved away. Eliot has begun to realise that some individuals cannot be happily rooted in a particular place, and the differences between the new world and the old cannot be blamed for this inability.

Silas Marner

Silas Marner (1861) is distinctive among Eliot's novels for its almost
total lack of reference to particular places and its great temporal
distance both from the time of Eliot's youth and from the present.
There is little reference to actual place or historical events in this novel;
the old world is very clearly separated from the new. We are told that
the events take place 'In the days when the spinning-wheels hummed
busily in the farmhouses',[28] but this is not so much an acknowledgement of change as an attempt to establish a different world, distant
enough for the reader to be unable to challenge Eliot's assumptions
about it. Geographical separateness is established by Eliot's
generalisations about the physical setting of the novel. All we are told
is that Raveloe 'lay in the rich central plain of what we are pleased to
call Merry England'.[29] Eliot is unwilling to let us know the name of the
town from which Silas has come, and only a few vague hints enable
us to suppose that he is from a Lancashire textile town. Change has
occurred within this remote and vague world, and Silas is confused
when he returns to his native town: 'Silas, bewildered by the changes
thirty years had brought over his native place, had stopped several
persons in succession to ask them the name of this town, that he might
be sure he was not under mistake about it.'[30] He finds a factory in place
of his old home, Lantern Yard, from which people are pouring 'as if
they'd been to chapel at this time o' day − a weekday noon!'[31]
   Silas has to undergo a significant change: he has to reject his roots
in his native town and grow new roots in Raveloe. When he first moves
in to Raveloe he lives in physical and social isolation from the village.
In contrast to Amos Barton's development of roots through the death
of his wife, Silas develops roots by his adoption of a child. Silas tries to
absorb the local culture for Eppie's benefit,

> as some man who has a precious plant to which he would give a
> nurturing home in new soil, thinks of the rain and sunshine, and all
> influences, in relation to his nursling, and asks industriously for all
> knowledge that will help him to satisfy the wants of the searching
> roots, or to guard leaf and bird from invading harm.[32]

His recognition that his old home has been torn down is the final event
which establishes his rootedness in Raveloe. Reporting his visit to

Lantern Yard to Dolly Winthrop, he says, 'The old place is all swep'
away . . . the little graveyard and everything. The old home's gone; I've
no home but this now.'[33]

If Eliot's own experiences at the time she wrote *Silas Marner* are
taken into account, the vagueness and distance of the temporal and
geographical settings of the novel become more significant. Eliot and
Lewes were living in Blandford Square, and at this time Eliot frequently
complained of 'physical weakness', 'feeble body', 'heavy eyes and
hands' and other symptoms of distress. She was very depressed and
Lewes' greatest efforts could not always help her to overcome this state.
Eliot herself blamed the 'London air' for her weakness and depression.
Haight writes that 'She had never liked life in London, had never ceased
to miss the country.'[34] Thus a peak in her reactions against London is
reflected in her writing by an idealisation of the past and of the values
of rural life — Silas violently rejects his native town, which he feels
offers him no welcome, and settles down into country life ever more
firmly: Eliot regrets the loss of rural roots which she feels she cannot
re-establish anywhere.

Romola

Romola (1863) bears very little resemblance, in terms of physical
and temporal settings, to the rest of Eliot's work. Set in late-fifteenth-
and early-sixteenth-century Florence, the novel overlaps with none of
Eliot's previous experience of time and place. Significantly, this novel
is also a product of the period during which she lived in Blandford
Square. Again Eliot asserts the importance of having roots in particular
places and communities; Tito, a foreigner who comes to upset Florence,
meets death at the hands of a man whose claims on his emotional
roots — the man who brought up Tito — Tito has refused to
acknowledge.

The geographical setting of this novel is stated very clearly, sometimes
very elaborately: thus,

> The Via de' Bardi, a street noted in the history of Florence, lies in
> Ottrarno, or that portion of the city which clothes the southern
> bank of the river. It extends from the Ponte Vecchio to the Piazza
> de' Mozzi at the head of the Ponte alle Grazie; its right hand line of

houses and walls being backed up by the rather steep ascent which
in the Fifteenth Century was known as the Hill of Bogoli, the
famous stone-quarry whence the city got its pavement . . . its
left-hand buildings flanking the river. . .[35]

On this occasion precise and firm rooting of the novel in a particular
time and place does not make up for Eliot's essentially superficial
knowledge and understanding of the place and time of which she is
speaking; the theme of roots is not treated in depth — but neither is
any other issue illuminated by Eliot's writing in this novel.

Felix Holt

At the end of 1863 Eliot and Lewes moved to a new house, the Priory,
at 21 North Bank; this, according to Haight, was 'a pretty, secluded
house, set far back from the street in a garden full of roses along the
Regent's Canal. . .'[36] Eliot lived here until her death, and in this more
congenial environment she again turned, in her writing, to the places of
her youth.

Felix Holt (1866), then, brings us back to the Midlands and to the
period of the Reform Act. The detailed introduction to the novel,
which describes many features of the parts of Warwickshire with which
she was most familiar, should not mislead us into thinking that Eliot is
writing a textbook of regional geography. For example, Mr Lyon lived
in Treby Magna — 'a small house, not quite so good as the parish
clerk's, adjoining the entry which led to the Chapel Yard'[37] — whereas
the original minister, Mr Francis Franklin, lived in a similar house in the
Chapel Yard of Cow Lane Baptist Church, Coventry.[38]

George Eliot's ideas about the need for roots are exemplified in this
novel by her characterisation of Harold Transome. Transome (having
been away from home for fifteen years) says, almost immediately after
his arrival at Transome Court, 'Gad, what fine oaks those are opposite!
Some of them must come down though.'[39] He is subjected to some
severe shocks in the course of the novel, two of which show him that
he is not the true heir to the estate which he has treated in such a
cavalier fashion; his rootlessness results in a lack of appropriate feelings
for the estate and Esther Lyon, a woman deeply rooted in the physical,
social, intellectual and emotional aspects of the area, is found to be the
true heiress.

Again in this novel Eliot recognises more fully that a distinction

between the past and the present which is equivalent to a distinction
between good and bad is an oversimplification of the truth. Change
again links the past to the present, as it did in 'Janet's Repentance', but
here Eliot seems even less keen on technological progress:

> The tube-journey can never lend much to picture and narrative; it
> is as barren as an exclamatory O! Whereas the happy outside
> passenger seated on the box from the dawn to the gloaming gathered
> enough stories of English life, enough of English labours in town
> and country, enough aspects of earth and sky, to make episodes for a
> modern Odyssey.[40]

The characters in *Felix Holt* react to their different pasts in different
ways. The conflict between different pasts becomes associated with the
reactions of different social classes to their pasts and futures. Felix tries
to mould the present in a way which is likely to make the future better;
and he uses his past experience constructively, sacrificing his own
prosperity to that of his class. Harold Transome's radicalism is designed,
by contrast, to help Transome retain his wealth and position in an
uncertain future. Coveney therefore suggests that there are two pasts
in this novel:

> There is the destructive past, associated with the tragic destinies of
> the aristocratic Transome family, which determines the present and
> destroys the future, and which exists in the present only as a
> 'sleepless memory that watches through all dreams'. And there is
> the other, the creative past, which nourishes the present with
> memory, harmonizing past and present in the sensibility, through
> the continuities and affections of the heart, which alone can
> establish the possibility of freedom and the morally chosen future.[41]

A character must have firmly established roots in his past as well as in a
place; Esther Lyon chooses the second of the two pasts which Coveney
suggests exist – and this carries with it implications of geographical
conservatism. Eliot sees the present as being in organic relationship to
the past, and it is this which enables her to give the penetrating account
of geographical change – through both time and space – with which
Felix Holt begins.

Middlemarch

In *Middlemarch* (1871–2) the regional emphasis of *Felix Holt* is
replaced by concentration on a small area centred on the town of
Middlemarch. Almost all the events of the novel take place in the part
of Warwickshire with which Eliot was most familiar – Coventry and the
parishes surrounding the town – although she had not visited the area
for seventeen years.

Eliot's understanding of the relative merits of the past and the
present, which I have suggested showed signs of growing in *Felix Holt*,
here reaches its greatest development. In *Middlemarch* Eliot can be
seen to recognise elements of four different worlds: a destructive past
and a creative past, a destructive present and a creative present (to
borrow Coveney's adjectives).

Bulstrode represents the destructive past; he has escaped from some
shady activities in London and on the Continent to become a highly
respected member of Middlemarch society. He is, however, an alien in
this environment and when, in conjunction with Lydgate, he tries to
alter the town, the destructive elements of his past reassert themselves
to cause his downfall. Lydgate comes to a new community which he
does not understand and begins to try to impose new medical practices
on it. Bulstrode says of their joint project:

> 'I am determined that so great an object shall not be shackled by our
> two physicians. Indeed, I am encouraged to consider your advent to
> this town as a gracious indication that a more manifest blessing is
> now to be awarded to my efforts. . .'[42]

In their attempt to impose new, town-bred ways, Lydgate and Bulstrode
move too quickly and without sufficient understanding of the place and
time in which they are acting, and, having failed to put down roots
gently over a long period, they are ousted, Lydgate chained to a girl,
Rosamond Vincy, who had thought that the local men were not good
enough for her. She represents the worst, most destructive qualities of
the present, being superficial, restless and rootless; and in marrying
Lydgate she marries a man who thinks that he can build a better future
all at once. Lydgate fails to use the creative elements of the past as a
foundation for the future.

Caleb Garth represents the idea that it is possible to establish creative
links between the past and the present. His understanding of his
physical and social environment is both great and sympathetic; his

attitude to tree-felling, for example, is creative, in great contrast to that of Harold Transome:

> The felling and lading of timber, and the huge trunk vibrating star-like in the distance along the highway, the crane at work on the wharf, the precision and variety of muscular effort whenever exact work had to be turned out — all these signs of his youth had acted on him as poetry without the aid of poets.[43]

Importantly, and almost uniquely in Eliot's novels, Garth is both knowledgeable about and happy with, those aspects of the Industrial Revolution which affect him. After a skirmish between his workers and some railway surveyors he remarks:

> 'Somebody told you the railroad was a bad thing. That was a lie. It may do a bit of harm here and there, to this and to that; and so does the sun in heaven. But the railway's a good thing.'[44]

While the local labouring classes are out of sympathy with their environment, because they cannot understand the changes in it, despite their local roots, the middle-class land agent has the breadth of vision to see his own, deeply loved, environment in relation to national forces. In this skirmish the narrowness of the old order has come into direct conflict with some of the enlightened aspects of the new world. Garth uses his past constructively and affectionately in conjunction with the more valuable aspects of the new technologies and knowledge available to him.

Ladislaw and Dorothea provide a different way of looking at the need to link the creative past and the creative present. Dorothea has only been at Tipton for a year when the book opens, but she has immediately involved herself in the geographical and social environment; 'she loved the fresh air and the various aspects of the country',[45] and she is shown drawing plans and persuading Chettam to build better cottages for his employees.[46]

Ladislaw, however, develops no roots in Middlemarch. Anderson maintains that he is 'so alien to Middlemarch that he cannot act on it directly', but he goes on to suggest, with no hint of irony, that this is partly a failure on the part of Middlemarch to take adequate account of change: 'the principles which prevail in the wider world outside Middlemarch cannot be articulated within it. The medium is too dense; it is not permeable from without.'[47] Although Ladislaw fails to develop physical roots, he is shown to have emotional, intellectual and moral roots which other characters − externally more rooted, because of long-term involvement in a particular place − seem to lack. He acquires emotional and moral roots by marrying Dorothea and entering Parliament, which enables him to use his accurately observed knowledge of many different people and places for the public good. Ladislaw's absorption of the past is contrasted with Casaubon's narrow-minded and detached observation of the past; Casaubon does not love the pictures he knows so much about, and he cannot order his voluminous notebooks on mythology into a unified whole. He does not see the present rooted in the past, he merely sees a number of discrete facts. This lack of love, Casaubon's failure to develop linkages between himself and his environment, is shown in his lack of attention to his house. The south and east of Lowick Manor are thus described:

> The grounds here were more confined, the flower-beds showed no very careful tendance, and large chunks of trees, chiefly of sombre years, had risen high, not ten yards from the windows. The building, of greenish stone, was in the old English style, not ugly, but small-windowed and melancholy-looking: the sort of house that must have children, many flowers, open windows, and little vistas of bright things, to make it seem a joyous home.[48]

Casaubon has made no attempt to make it into a 'joyous home'. Firmly rooted in this particular physical environment, Casaubon has no emotional, social or intellectual roots of any depth. In contrast, although Dorothea and Ladislaw have few long-term physical roots, they are justified, in Eliot's eyes, by the depth of their emotional, intellectual and social roots. In this novel physical roots are less important than emotional, affective and moral roots; Eliot, in Anderson's words, transposes the 'natural into the moral and psychological'.[49]

Eliot had not visited the area in which *Middlemarch* is set for seventeen years, and the novel contains very little physical description. At the beginning of this chapter I suggested several reasons for the decreasing amounts of topographical description in Eliot's novels; here it might be possible to attribute this, at least in part, to Eliot's insistence that non-physical roots are more important than physical ones. The physical is, in general in this novel, less important than the social, moral and emotional; therefore physical description is less important. In fact, Anderson writes:

there are only two fully realized natural landscapes, Lowick Manor and Stone Court, and in these cases the landscape is realized by an individual whose situation and interests make him aware of an external world at that particular moment. For the most part we may characterize the book's use of the physical world by referring to George Eliot's own sense of Warwickshire as a physical locale which has been wholly humanized and to the Reverend Mr. Cadwallader's half-serious remark that it is a very good quality in a man to have a trout stream.[50]

Eliot finds the climate and soil most suitable for the growth of roots in Ladislaw's wide vision: they establish a new world, freed from the physical and social narrowness of a small provincial town: 'All their vision, all their thought of each other had been as in a world apart, where the sunshine fell on tall white lilies, where no evil lurked, and no other soul entered.'[51]

Daniel Deronda

Eliot finally frees herself from her bondage to the past in *Daniel Deronda* (1874–6). Here she seems -- at first sight – to have worked out her own feelings of rootlessness and unhappy detachment from the places and values of her youth by continuing the idea of the superiority of moral, emotional and social roots over physical roots. She does this by setting her novel in, approximately, the present and in places distant from the Midlands (in Wessex, Leubronn and Genoa). The location of the novel is again difficult to ascertain precisely, as we are told only about 'all classes within a certain circuit of Wancester',[52] 'a few people in a corner of Wessex',[53] and 'the admiration for the handsome Miss Harleth, extending perhaps over thirty square miles in a part of Wessex'.[54]

Yet when we begin to look more closely at the novel and look beyond the evidence which shows us the novel's apparent detachment from the times and places which have hitherto dominated her novels, we soon come to realise that Eliot's concern for roots and rootlessness is as insistently present here as it has always been.

Eliot does not seem to have found lasting satisfaction in the solution to the problem of rootlessness among the upper classes presented in *Middlemarch*. Gwendolen Harleth lacks physical roots; the quotation at the beginning of this chapter presents Eliot's views about Gwendolen's rootlessness. The quotation is a dogmatic statement of Eliot's view that physical rootedness is necessary as a fertile soil for the growth of emotional, social and intellectual roots. Gwendolen has to undergo a moral transfiguration under the tutelage of Deronda, the messianic Jew; but at the end of the novel Gwendolen has still not found her 'roots' in any particular physical environment, although her moral roots are much deeper. We still do not know whether they are deep enough to survive the loss of Deronda, on whom she has grown emotionally dependent.

Deronda himself has deep roots in his home environment: sitting on his windowsill

> he could see the rain gradually subsiding with gleams through the parting clouds which lit up a great reach of the park, where the old oaks stood apart from each other, and the bordering wood was pierced with a green glade which met the eastern sky. This was a scene which had always been part of his home . . . and his ardent clinging nature had appropriated it all with affection.[55]

Deronda uproots himself from this environment; he more or less decides that he wishes he was a Jew, and then discovers that he is Jewish by birth. He takes up the cry of his dying friend Mordecai, marries Mordecai's sister, and sets off for the Middle East. Here is Mordecai's call:

> In the multitude of the ignorant on three continents who observe our rites and make confession of the divine Unity, the soul of Judaism is not dead. Revive the organic centre: let the unity of Israel which has made the growth and form of its religion to be an outward reality. Looking towards a land and a polity, our dispersed people in all the ends of the earth may share the dignity of a national life which has a voice among the peoples of the East and the West — which will plant the wisdom and skill of our race so that it may be, as of old, a medium of transmission and understanding.[56]

Ultimately, people need an 'organic centre'; and that centre must have some sort of physical expression. For meaningful and useful existence a race, in this example, requires a land. Thus, in her last novel, George Eliot is at the same point as she was in 'Amos Barton' — emotional, moral and social development is ultimately dependent on having an adequate depth of roots in 'some spot of a native land'.

Notes

1. G. Eliot, *Daniel Deronda*, 1874–6 (Penguin, Harmondsworth, 1967), p. 50.
2. G. Eliot, *Middlemarch*, 1871–2 (Pan Books, London, 1973), p. 92.
3. G. S. Haight, *George Eliot: a biography* (Oxford University Press, London, 1968), pp. 8–175.
4. H. Auster, *Local Habitations: Regionalism in the Early Novels of George Eliot* (Harvard University Press, Cambridge, Mass., 1970), p. 61.
5. G. Eliot, *Scenes of Clerical Life*, 1857 (Penguin, Harmondsworth, 1973), pp. 252–3.
6. Ibid., p. 253.
7. Ibid., p. 412.
8. Haight, *George Eliot*, p. 1.
9. Eliot, *Scenes*, p. 158.
10. Eliot, *Scenes*, p. 113.
11. Eliot, *Scenes*, pp. 112–13.
12. Quoted in Haight, *George Eliot*, p. 68.
13. Auster, *Local Habitations*, p. 115.
14. G. Eliot, *Adam Bede*, 1859 (Dent, London, 1960, Everyman edition), p. 151.
15. Ibid., p. 356.
16. Ibid., p. 337.
17. Ibid., p. 88.
18. G. Eliot, *The Mill On The Floss*, 1860 (Oxford University Press, London, 1903), p. 302.
19. Auster, *Local Habitations*, p. 173.
20. Ibid., p. 164.
21. Eliot, *The Mill*, p. 458.
22. Ibid., p. 18.
23. Ibid., p. 607.
24. Auster, *Local Habitations*, p. 163.
25. Haight, *George Eliot*, p. 305.
26. Ibid., p. 302.
27. G. S. Haight (ed.), *A Century of George Eliot Criticism* (Methuen, London, 1966), p. 339.
28. G. Eliot, *Silas Marner*, 1861 (Dent, London, 1906, Everyman edition), p. 3.
29. Ibid., p. 4.
30. Ibid., p. 255.
31. Ibid., p. 251.
32. Ibid., p. 210.
33. Ibid., p. 258.
34. Haight, *George Eliot*, p. 338.
35. G. Eliot, *Romola*, 1862–3 (Nesbit, London, 1905), p. 35.
36. Haight, *George Eliot*, p. 371.
37. G. Eliot, *Felix Holt*, 1866 (Penguin, Harmondsworth, 1972), p. 131.
38. J. W. Cross, *George Eliot's Life as Related in Her Letters and Journals* (Harper and Brothers, New York, 1885), vol. 1, p. 18.
39. Eliot, *Felix Holt*, p. 95.
40. Ibid., p. 75.
41. P. Coveney, Introduction to *Felix Holt*, p. 9.
42. Eliot, *Middlemarch*, p. 111.
43. Ibid., p. 288.
44. Ibid., p. 510.

45. Ibid., p. 4.
46. Ibid., p. 503.
47. In Haight, *A Century*, p. 317.
48. Eliot, *Middlemarch*, p. 63.
49. In Haight, *A Century*, p. 317.
50. Ibid., p. 317.
51. Ibid., p. 735.
52. Eliot, *Daniel Deronda*, p. 122.
53. Ibid.
54. Ibid., p. 130.
55. Ibid., p. 208.
56. Ibid., p. 592.

7 ON YEARNING FOR HOME: AN EPISTEMOLOGICAL VIEW OF ONTOLOGICAL TRANSFORMATIONS

Gunnar Olsson

> I would rather be a man of paradoxes than a man of
> prejudices — Jean-Jacques Rousseau

Longer back than I wish to remember, I promised a piece for the collage
of this volume. At first, I imagined that fulfilling the obligation would
be easy, for all I intended was to relate some personal experiences. But
then, gradually, I felt how deeply my words had committed me. I tried
to formulate proper expressions. I tried again. I tried again and again
only to find that my pen left nothing but invisible marks on white
sheets of paper. In the acts of zero degree writing, one deadline was put
on top of another.

It now appears that the furtility of my writing itself illustrates a
major point I was aiming to make. This is that as time and space curl in
on each other, so they curl in on themselves. At the centre are issues of
self-reference. Is it possible, for instance, to approach a topic as crucial
as yearning for home in a mode which is not logically self-referential?
Is it possible to write anything worthwhile in a language which is not at
the same time a denial and an affirmation both of what it tries to
picture and of the picturing itself? Is it possible to reach for truth
without confronting the paradox of Epimedes, the Cretan, who coined
the immortal statement 'All Cretans are liars'?

i think not

Part of the reason is in Gödel's Incompleteness Theorem in which
numbers served as codes for statements about numbers. It is now clear
that Gödel's attempt to write *in* the language he was writing *about* blew
the world apart, for its polytheistic implication is that every mode of
expression is incomplete. More technically, Gödel demonstrated that it
is possible to demonstrate that there are true statements in every system
of thought which are not demonstrable within that system; provability
proved itself to be a weaker notion than truth. It is exactly the same
problem of logical types which is at the heart also of the hermetic

tradition of modern literature. Stéphan Mallarmé, James Joyce, Samuel Beckett and Arno Schmidt come to my mind. J. S. Bach and M. C. Escher could come to the mind of others.

I share the artists' obsession with the interface of logical types and I am attracted by their attempts to move freely between the levels. In what follows, I shall seek entrance into the abyss-filled territory they call their own. I do so in the hope of catching an understanding of what it means to yearn for home. I am driven into the boundless realm of poetry by the recognition that it is impossible to give a stable and for ever valid definition of what yearning is. The reason is that the truth of yearning is in the yearning itself, not in the things and relations the yearning is for. There is nevertheless the possibility of catching a glimpse of the untouchable by performing a striptease on some of the symbols in which the feeling has taken concrete, albeit temporary, form. And yet, even though I can touch the particular of a symbol, its meaning always remains evasive. It follows that I shall be stretching toward a relation I can experience but in fact neither touch nor see. The concept of yearning belongs to one logical type. The symbol of home lies on another.

My method is dialectical and grows out of the experience that understanding can never be direct. Insight is instead gained through negation, hence through others; the 'I' can become conscious of itself only through the 'me'. The epistemological lesson is that analysis must not be limited to the phenomenon seemingly at issue. It must instead be contextual and try to illuminate the study object both through its immediate opposite and through its counterparts on other levels of logical types. In addition, it is important to acknowledge that meaning is an intersubjective relation, for experience is not meaning until communicated and thereby destroyed.

This is the dilemma, because the meaning a sender imputes to the symbol he emits may not be the same as the meaning the receiver detects. Since communication without metaphor is impossible, perfect translation is impossible as well. From the obedient viewpoint of social cohesion this is a serious threat. In the creative perspective of individual ambiguity, on the other hand, it offers the only way out. But how do I ground my representation when my ambition is to criticise and thereby repressent society altogether? How do I make others aware of my rejection of current values when the same norms serve to legitimise a society which does not correspond to them? How do I tell meaningful truths about a world whose very nature it is to be a lie? Epimedes, again!

My present ambition is to communicate the meaning of yearning. In
one sense, I am foredoomed, for yearning is a relation between myself
~~as I think I once was~~ and something I lack ~~in what I presently take to be.~~
The anchoring of my thought can therefore not be in what I am
thinking about but must instead be in the thought itself. This reraises
Gödel's question of whether such a feat is possible. The answer:
'Completely, no; incompletely, yes.'

The challenge is in understanding and capturing the double dialectic
of symbol and meaning. One danger here lies in thingifying the reasoning
and thereby equate the yearning with what I am yearning for. Another
temptation is to spiritualise the matter and equate what I am yearning
for with the yearning. Use and mention must not be confused, though,
because the meaning of a name is not the bearer of the name and the
bearer of the name is not the meaning of the name. Meaning is instead
in use and the approach is to materialise spirit and spiritualise matter.
It follows that meaning is contextual just as both Marx and Wittgenstein
claimed it to be. What is at stake is the Leibnizian principle of
substitution, for it all depends on who says what to whom where and
when.

Since meaning is contextbound, I should stress that I take
epistemology to be my context, sense of place my topic and philosophical literature my major data base. It follows that when I argue for
the dialectical method, I do not necessarily assume that the world
consists of hard opposites. I do assume, however, that understanding
the world always begins in definitions of firm categories, proceeds first
to their negation and then to a level at which affirmation and denial
cancel out.

Even though the universe comes into being with the drawing of a
distinction, that act is not final but endless; knowledge knows only
provisional boundaries. What is left at each moment is consequently
nothing but traces of what went before. We begin by setting limits, for
understanding is in the act of transcending them just as Wittgenstein
constructed his ladders in order to throw them away. If I want to
understand notions like regional identity and home, I must therefore
also understand historical change and exile. In such a perspective,
present is merely an instance between past and future just as home is
nothing until lost. And yet, it is the conjunction of that merely nothing
which the present article chases around.

To yearn for home is to be entangled in the connections between the
logical types of thing and relation, word and object, matter and mind,

internal and external. It has been the aim of traditional social science to capture these connections in a net of causal analysis. The catch has frequently yielded accurate predictions of aggregate behaviour, sometimes even valid explanations. Often, however, the lure of positivism has led the analysist into the double trap of logical exactness and misplaced concreteness. But nothing occurs in reality which strictly corresponds to logic, just as nothing occurs in theory which strictly corresponds to reality. As a snapping consequence, many now argue that the specification of causal relationships should be complemented by attempts to unravel symbolic relations. Under the influence of such tunes, parts of the social sciences have inched away from causal analysis and moved closer to the emerging paradigm of semiotics. In the process, the concern with social engineering techniques of manipulative power has diminished in favour of a search for emphathic understanding of man's efforts to understand himself and his culture. But even though one thereby hopes to reveal some hidden aspects of domination, hope must not be mistaken for truth; the will to control shifts masks quicker than it is within the power of the participant observer to observe.

An important forerunner to the new tradition of semiotics was Ernst Cassirer, who devoted his major works to the philosophy of symbolic forms. I share his fundamental idea that there is no meaning in physical expressions taken in isolation. Sounds become words, movements gestures, clothes signals, only in the context of complicated communication processes in which meaning is extracted from the systematic differences among the constitutive elements of the interpreted text. It is in this dance of opposites that things, thoughts and actions turn themselves into symbols. It is in that realm of semiotics that human eyes cannot only see but ask and answer questions as well. It is there that we experience ourselves as lonely individuals and as members of social crowds at the same time.

The point I have been working up to is that the human world is the only world we can have meaningful contact with. But this world of ours is not limited to externally defined existing physical things. In addition, it includes internally sensed objects. In the former case we are concerned with signification, which is essentially digital, while in the latter our interest is in meaning, which is essentially annalogous. Thus, there are not only wooden constructions, but chairs and tables; not only movements of arms and legs, but friendly and unfriendly acts; not only vibrations in the vocal chord, but promises to keep; not only lips to lips, but sincere and deceptive kisses. To understand the social and cultural interaction of individuals and groups is therefore not only to describe atomistic and independently observable signs. In addition, it is to decode the symbolic structures through which our objects and actions preserve and create a universe of human community. One result of these decodings is that physical objects turn out to be material expressions of social relations just as untouchables become spritual expressions of material relations. It thus appears that social scientific truth must be of the coherence type.

It is an integral part of all coherence theories that the elements of a text are not interpretable in isolation. Meaning is instead in the internal relations which keep the parts and their opposites together not only when they are mentioned but also when they are used. In turn, these relations are tied in with other relations into a kind of self-referential spiralling *ad infinitum.* Bankers and beggars become complements of each other, for it is only the rich who can afford poverty. But if I liken a woman to the moon, then my parable lacks interpretation until it is taken as a mythical manifestation of the creative opposition between woman and man, moon and sun, night and day, 0 and 1. That context is of course cultural just as the internal relations of culture themselves are cultural. To feel at home outside one's own circle is therefore impossible, for it is in the nature of internal relations that if one link is broken, then the whole world tumbles down. Noone can put Humpty Dumpty together again; the broken egg yields not a bird but an omelette.

It should now be obvious that I take the interplay of physical material (existing objects) and spiritual meaning (subsisting objects) to be internal of all symbols. Within such a context, it is easy to appreciate T. S. Eliot's argument that symbols are objective correlates of human feelings. And yet, there is a persistent drift toward thingification and objectification, significantly in the name of communication itself. To be alien is in fact to live in the rift between alternative and equally valued sets of internal relations. When I obey one rule, I am bound to break the commands of another; you are damned if you do and damned if you don't.

There are lessons to draw. One is that if society is becoming increasingly autonomous of its makers, then the opportunities for affecting your own situation decrease. But linked with this recognition is the suspicion that the captivity of modern man is one he somehow brought on to himself. Perhaps he has even institutionalised a potentially pathogenic world of double binds. Such a world of universal domination is likely to go mad if those in power are allowed to use double bind as a weapon of control. The prospects are uncertain,

because it is a characteristic of today's world of economics and ecology
that it can afford neither to cease growing nor to continue growing.

Alienation! Predicament! Double bind!

The concluding twist is crucial, for alienation, predicament and
double bind is at the root of social crisis, tragedy and schizophrenia.
This suggests that those who cannot unconsciously live with the truth
of Gödel's Incompleteness Theorem are banned from normal life, for
to be normal is to live in double bind and be rewarded for the failure of
recognising it. Self-reference is at the core. He who cannot straddle the
line between the complicity of the inside and the rebellion of the
outside is defined away as neither belonging nor not-belonging. He
whose communication is devoid of meta-rules becomes a category in
himself. And so it is that the tragic hero is stuck in predicament, the
schizophrenic in double bind, social crisis in alienation. To be stuck is
in this context to be confined to one level alone in the theory of types.
To be stuck is no joke, for to joke is to play with logical types.

What does it *now* mean to yearn for home? Jokingly serious, seriously
joking?

Joker trumps the trumps! Follow suit!

To yearn for home is to experience double bind. It is to be torn between
irreconcilable identities, sometimes enjoying the ~~illusory~~ freedom of
swinging with the wind, sometimes missing the ~~real~~ subjugation of being
fettered to the ground. As before, the challenge is in not confusing the
yearning with what the yearning is for. The difficulty is contextual, for
in the reified world of modern culture it is natural to get stuck on the
thing-level of logical types; every mode of analysis hates the idea of
becoming and loves the fact of being. To understand such a context is
to reveal the hidden internal relations of the symbols in which the lack
momentarily expresses itself.

It is part of my epistemological stance that insiders experience and
outsiders understand; whereas experience is confined to one logical
type at the time, understanding is in the act of crossing categorial
boundaries. For this reason, it is not surprising that the most
penetrating accounts of home stem from people away: August
Strindberg, Henry James, James Joyce, Marc Chagall, Witold
Gombrowicz, Vladimir Nabokov. . . From the perspective of double

bind it is equally interesting that the yearning often is rendered as a
return to physical, indeed earthy, objects.

In Swedish literature, Verner von Heidenstam (1859–1940) provides
the best example of the drifts I have mentioned. Already at the age of
seventeen, ill health forced him into eleven years of wandering through
the warmer climates of Southern Europe and the Middle East. Far away
he confronted other cultures, searching there for the riches of Mogul's
ring. He never experienced them, perhaps because he looked too hard
and understood too well – *Hans Alienus* became the fitting title of his
autobiographical novel. And in his early 'Ensamhetens tankar' he wailed

> I yearn for home.
> Since eight long years I yearn for home.
> In deepest sleep I felt the longing.
> I yearn for home. I yearn wherever I roam
> – but not for people. I yearn the ground,
> I yearn the stones where once I played.[1]

Heidenstam yearns for home. Purified in the flames of time, the
feeling brings him closer first to the self-reference of yearning and then
to himself. As it does, the groping seeks its symbols not in ambiguous
social relations with people but in the unyielding rocks of the ground
he grew out of.

Having spent as many years in exile as Heidenstam, I share his
experience of ontological transformations; as amorphous people recede
into shadowy memory, hard objects assume their place in the sun.
Human touch is fading away. But what I remember better than I know
is the roundness of the stone in my hand before I threw it and the
sharpness of the grass in which I was hiding from the chasers. Not
people but stones and grass. Not the shifting nature of humans but the
stable humaneness of nature. You yearn back to the earth you were
taken out of, your mind laid to rest in the self-reference of dusty
ground. It is when I reach this modicum of understanding that I begin
to suspect that the collective insights of mythology, epistemology,
ontology, psychology, semiology may all be reaching for the same:

> –ology–
> –logic–
> –logos–
> –reason/word–
> –treason/world–
> –letters crossing into new realities–

or, in other marks: reality is in thought, thought in logic, logic in ethics, ethics in words, words in ontology, ontology in epistemology, epistemology in action, action in conjunction, conjunction in dream, dream in reality.

Dreams and realities, all in the same sign in the sky. And so it is that when Heidenstam felt the longing in his deepest sleep it was not for flashing people but for murmuring stones. Such ontological transformations lie indeed at the heart of all surrealism. Thus we often populate the same dream with persons from different times and places; grand fathers, parents, children, friend and foe all mingle together. What appears more rare is to have the dream world furnished with physical objects from incongruent environments; the Ann Arbor house is not set in the woods of Värmland and the pebbles from the Tasmanian Sea are not on the floor of the Hongkong barroom; even Breton seldomly let Nadja out of Paris.

If dreams are what I believe — keys to the collective unconscious — then I take my biased sample to exemplify the close ties of meaning and thingification. Moreover, I sense the return of old myths as I see the present exposing itself in the double mirror of ~~the lacking~~ past and future. When I thus admit my cultural heritage, then I begin to feel the living context in which *The Odyssey* became *Ulysses* to be Found Again in *Finnegan*. Here Comes Everybody's Always Longing Pack.

And so it comes about that yearning for home fixes itself on seemingly trivial things. The ambiguous seeks a stable form as dreams get stoned into objective correlates. When the longing and its symbols meet on equal terms and common ground, meaning turns to marble, marble to meaning.

Self-reference is the name of the game. It goes on for ever, for in the end noone can win. To yearn for home is merely one among many attempts to close and jump out of the circle. It will not succeed because it is part of the human condition to live alienation and double bind. Fear of crossing the existential and social boundary into death and silence is implicated. Indeed it is mainly in the symbols of death that we afford to acknowledge the inevitability of self-reference. Heidenstam's stones serve as mediators between ambiguous people and certain dust.

Dust to dust. Noone will grasp the meaning of that symbol even though the poets seem closer than the social scientists. But noone escapes from the ghost of Epimedes, now or then, home or away. Merely nothing.

As if to top it off, Evert Taube — Sweden's most popular troubadour —
sang about the love who left:

> Jag lade mig på marken,
> jag sade hennes namn,
> jag tryckte mig mot jorden
> som var den hennes barm.
>
> Jag var så trött av kärlek
> jag var så trött av sorg.[2]

Notes

1. Jag längtar hem sen åtta långa år.
   I själva sömnen har jag längtan känt.
   Jag längtar hem. Jag längtar var jag går
   – men ej till människor. Jag längtar marken
   jag längtar stenarna där barn jag lekt.
2. I put myself onto the ground
   I said out loud her name
   I pressed my body to the earth
   as if it were her same.

   .

   I was so worn from sorrow
   I was so worn from love.

8 LITERATURE AND THE FASHIONING OF TOURIST TASTE

Peter T. Newby

The use of place as a setting for a story or as a source for creative endeavour is well established in English literature. In some instances there is no doubt that writers have helped establish a tourist locale; Daiches and Flower provide several examples in their recent study of literary landscapes.[1] Nor was it, or is it, unknown for authors to attempt to capitalise on the fashion of place. The concern of this chapter, however, is not so much with the potential of literature to create and endorse fashionable places (although certainly this is an aspect which has to be considered) but rather with its ability to influence styles of tourism. This theme will be illustrated with reference to tourism in two contrasted areas, the English Lake District and the Mediterranean.

Lake District Tourism

The lakes and mountains of Britain were 'discovered' towards the end of the eighteenth century when political changes on the Continent restricted travel opportunities for those who might have undertaken the Grand Tour. At this time also, attitudes as to what constituted attractive landscape were changing. These changes were apparent in, and broadcast through, the works of the Romantic poets and novelists. The Romantic Movement saw beauty in the organisation of natural forces rather than in an intellectual order; it permitted the expression of feeling and emotion, and praised the way man and nature coexisted. Wordsworth in particular epitomised this new philosophy, exerting considerable influence on British literary and landscape taste. His poetry paved the way for the development of a new tourist style because of its emphasis upon nature and man's response to, and feeling for, the natural world. It represented a philosophy that emotions themselves were not unworthy, and that the natural world could produce an emotional response. It was the acceptance of these ideas that led to the growth in scenic tourism. His appreciation of form involved an understanding of the underlying qualities and relationships of nature until it provided a morality with which to assess the world. It is not the morality that is, however, our prime concern, for the judgements which his view of nature led him to make sometimes conflicted with the realities of tourism. More important for the development of tourism was his ability to describe landscape and evaluate the impact of landscape and nature in emotional terms, and in this way to communicate a new taste and philosophy. Lacey has criticised him for depending 'for the continuance of his imaginative life on the forms of nature rather than on the interfusing spirit of nature',[2] and according to literary standards this judgement may be valid, but had Wordsworth sought stimulus in the 'spirit of nature', there would have been fewer word portraits to stimulate the imaginations of tourists and would-be tourists.

Wordsworth's impact in the late eighteenth and nineteenth centuries was as a poet who expressed the forms and qualities of nature, and in showing how man's life and attitudes were bound up with places and objects in the natural world. His many statements of the reactions and emotions that nature could induce gave credibility to the view that satisfaction could be sought in other than rational endeavours, and it was the progressive acceptance of this point of view that was necessary for the growth of scenic tourism. Wordsworth, however, did more than help create the intellectual climate for a new type of tourism, he also identified the locations which the tourists should visit. He did this in two ways: first, through the use of place in his verse, and secondly through his authorship of one of the most popular nineteenth-century guides to the Lake District.

The use of place in his verse is an extension of his concern with nature. Many of his poems clearly explore what today we would call the 'spirit of place' — the significance of locations for the attitudes, emotions and activities of man. Nowhere is his perception of the importance of place more evident than in the series of poems concerned with the naming of places. In fact, Wordsworth's philosophy of place is particularly clear in these poems in that he demonstrates, not only that place has an impact on man, but also that through man place becomes alive and significant. His use of place was important for the tourist industry because it demonstrated where the experiences that gave rise to his philosophy and new standards of taste could also be experienced by others. In Britain Wordsworth uses places as far afield as Salisbury Plain and the Quantocks in the South and the Inner Hebrides in the North, but he drew most extensively, of course, upon the Lake District.

In his narrative poems Wordsworth describes landscape and uses place as a backdrop to the action. The scramble up Green-head Ghyll to the site of a sheep fold provides the story told in 'Michael' with an authenticity that could only but appeal to a tourist. But perhaps of most significance for a person wishing to experience the pleasures of landscape were his extensive landscape descriptions. On occasions his verse provides us with a topography of an area:

> That ancient Woman seated on Helmcrag
> Was ready with her cavern; Hammarscar,
> And the tall Steep of Silver-how, sent forth
> A noise of laughter; southern Loughrigg heard,
> And Fairfield answered with a mountain tone;
> Helvellyn far into the clear blue sky
> Carried the Lady's voice, — old Skiddaw blew
> His speaking-trumpet; — back out of the clouds
> Of Glaramara southward came the voice;
> And Kirkstone tossed it from his misty head.[3]

The importance of landscape description to Wordsworth's philosophy is apparent from the extensive use he made of it in his poetry. That he wants to convince others of the qualities and significance of landscape is apparent in these lines from *The River Duddon* sonnets:

> How shall I paint thee? — Be this naked stone
> My seat, while I give way to such intent;
> Pleased could my verse, a speaking monument,
> Make to the eyes of men thy features known.[4]

Wordsworth's knowledge of the Lake District was extensive, yet his use of scenes and locations in his verse was restricted, being clearly concentrated around Rydal water, near where he lived at Allan Bank and Rydal Mount. He was born in Cockermouth in 1770 and the years until 1787 were spent in and around the Lake District. The next twelve years were spent away from the North-west, but it was, nevertheless, an important period in his life in which his attitude to nature crystallised and his association with Coleridge flourished. In 1799, following a walking holiday with Coleridge in the Lake District, the Wordsworths settled at Dove Cottage in Grasmere. He lived in the Lake District until his death in 1850, moving to Allan Bank in 1809 and Rydal Mount in 1813.

His associations with Coleridge and Southey and the name by which they were collectively known — the Lakes poets — served to highlight the area.[5] But it was not just the verse that directed popular attention to the new philosophy and the landscapes associated with it, for reviews and excerpts were probably disseminated at least as widely as the works themselves.[6]

The increasing popularity of the Lake District for tourists is apparent from an event of 1809. In that year Wordsworth agreed to write a description of the Lake District as an introduction to a series of sketches by the Reverend Joseph Wilkinson. (This was one of two such volumes in preparation at the time; William Gilpin, a friend of Wordsworth, preparing the other.) The sketches and Wordsworth's description were published in 1810, and Wilkinson republished it in 1821. Wordsworth, however, had already published a slightly amended version of the description together with *The River Duddon* sonnets in 1820. It was so successful that it was revised in 1822, 1823 and 1835.[7] After 1835 Wordsworth was unwilling to amend his guide, but he allowed his publisher John Hudson to use his material in a series of guides dating from 1842.[8] While the print-runs on Wordsworth's guides reveal the increasing popularity of the Lake District as a tourist area (the 1822 edition was for 500 copies, the 1823 for 1,000, and the 1835 for 1,500), it is clear that the age of mass tourism had not yet arrived.

Wordsworth's guide is much more than a description of routes and objects. He sets out to dissect the landscape and to explore the natural and man-made aspects in an attempt to understand its character and quality. That he was not writing another tourist tract is seen from the first sentence:

> In preparing this Manual, it was the Author's principal wish to furnish a Guide or Companion for the *Minds* of Persons of taste, and feeling for Landscape, who might be inclined to explore the District of the Lakes with that degree of attention to which its beauty may fairly lay claim.[9]

His analysis is systematic, comparative and historical, and his overall object is to evaluate the aesthetic quality of landscape and to set out guidelines for conserving landscape quality. It is this demonstration of aesthetic quality, particularly in comparative terms, that helped establish the tourist fashion for the area. Wordsworth's canons of criticism are landscape scale (in which there is little merit in magnitude *per se*), the relationship of landscape components and a preference for variety. Perhaps inevitably, therefore, the Lake District appears more attractive than the Scottish Lochs, North Wales, or even the Swiss Alps with which it is compared. How many seasoned tourists would have

agreed with this comparative judgement cannot be estimated, but
there is no doubt that through his poetry and prose Wordsworth drew
a picture which must have been very appealing to tourists. His attitude
to tourists, however, might appear to be rather cynical for someone
who had written a well-received and highly influential guide. In 'The
Brothers', for example, 'the homely Priest of Ennerdale' gives the
following description of early tourism:

> These Tourists, heaven preserve us! needs must live
> A profitable life: some glance along,
> Rapid and gay, as if the earth were air,
> And they were butterflies to wheel about
> Long as the summer lasted: some, as wise,
> Perched on the forehead of a jutting crag,
> Pencil in hand and book upon the knee,
> Will look and scribble, scribble on and look,
> Until a man might travel twelve stout miles,
> Or reap an acre of his neighbour's corn.[10]

Wordsworth's guide was intended for just such people, people who
wanted to enhance their appreciation of landscape and views, people
for whom scenery was an intellectual challenge, not an immediate
reward – in other words, 'people of taste' for whom tourism represented
as much a voyage of the mind as one of the body. He foresaw, however,
that popularisation would have attendant problems, not the least of
which would be that 'people of taste' would no longer constitute the
majority of tourists. His opinion is manifested in the letters he wrote to
the *Morning Post* concerning a proposed railway that would join
Windermere and Kendal to the main Lancaster–Carlisle line. He clearly
drew a distinction between the tourist, a person of taste, and the
railway visitor who would be unable to appreciate the landscape. An
influx of the latter type would threaten what Wordsworth called 'the
staple of the district . . . its beauty and its character of seclusion and
retirement'.[11] Wordsworth's elitist view stemmed from a feeling that
his own understanding, developed as it had been over a lifetime (he was
seventy-five when he wrote the letters), would be trivialised by an
influx of 'uneducated' visitors. Wordsworth's arguments were to no
avail. The railway was built between 1846 and 1847 and the number of
visitors to the Lake District grew. Passenger returns from the
Cockermouth, Keswick and Penrith Railway, opened in 1865, show a
pattern of growth to reach a peak of nearly half a million in 1907. The

bulk of this number represented third-class traffic, which was encouraged by the railway fare policy with the introduction of the 'excursion class' and with higher second-class fares subsidising an increasing proportion of third-class (nine-tenths of the total by 1880). The exclusive character of the area was affected accordingly – the tourist fashion, which Wordsworth had helped to initiate, had become an accepted mode for the working classes. The fashion leaders, meanwhile, had moved on to develop new areas and new styles.

The Mediterranean and the Tourist Season

Some of the leaders of fashion that had deserted the Lake District may well have found their way to the French Riviera, where towns such as Nice, Cannes and Menton had emerged as winter health resorts by the mid-nineteenth century. By the early twentieth century, however, railways in turn were beginning to destroy the exclusiveness of these resorts. The resultant diffusion of the Riviera winter season through the social hierarchy, however, was augmented after the First World War by an event which was to stimulate a popularity in the Mediterranean *summer* season.

In 1924 Scott Fitzgerald and his wife moved to the Riviera. He was already well known as a novelist and writer of short stories,[12] while his ability to spend money more quickly than he earned it meant that he and his wife were well-established figures on the New York social scene. Their move to the Mediterranean, even though it was due partly to financial pressures, undoubtedly influenced where New York society chose to spend the summer. They themselves spent the summers of 1924, 1925, 1926 and 1929 on the Riviera.

The publication of *Tender is the Night* in 1934 was a decided encouragement to the growth of a summer Mediterranean season. At the opening of the novel the author remarks that the French Riviera had lately 'become a summer resort of notable and fashionable people', whereas 'a decade ago it was almost deserted after its English clientele went north in April'.[13] This newness of the summer season is illustrated in the following early exchange between Rosemary and Abe North in the same novel:

'Do you like it here – this place?'
'They have to like it,' said Abe North slowly. 'They invented it.'
He turned his noble head slowly so that his eyes rested with

tenderness and affection on the two Divers.

'Oh, did you?'

'This is only the second season that the hotel's been open in summer,' Nicole explained. 'We persuaded Gausse to keep on a cook and a *garçon* and a *chausseur* – it paid its way and this year it's doing even better.'[14]

The success of the summer season is then explained by Dick Diver:

'The theory is . . . that all the northern places, like Deauville, were picked out by Russians and English who don't mind the cold, while half of us Americans came from tropical climates – that's why we're beginning to come here.'[15]

Perhaps one of the most significant features of Fitzgerald's influence on the development of the summer season is that it was achieved with little landscape description – certainly far less than that characterising the work of Hardy, Scott or Wordsworth. His influence is due to the fact that he captured and correctly interpreted social change, both in his novel and in his own lifestyle, and in so doing helped to create a new tourist fashion. He saw that the war and its economic consequences created a desire for holidays that were a release from the problems of the times, and that the intellectual appreciation of landscape had given way to the superficial delights of sun and sand as the motivating force for the fashion leaders. Once the isolationism of the United States had been destroyed by the First World War, the opportunity for large-scale American foreign tourism existed.

Scott Fitzgerald, by his action of living on the Riviera and through publication of *Tender is the Night*, showed Americans what was in vogue. The summer season was not, however, solely an American affair for long, for the influence of American society was such that fashion-conscious Europeans soon followed suit. Thus, what Wordsworth did for lakes and mountains, Scott Fitzgerald did for summer-sun holidays. Both preached a style that found favour with those able to travel; both portrayed their styles in a particular location and, in both cases, what started out as fashion creating and reflecting a change in taste, became established tourist tradition as a consequence of a coincidental and fortunate expansion of transport opportunities. In both periods personal incomes and periods of paid holidays were increasing, but it was transport developments that really created the opportunity. In Wordsworth's time railways provided the means that permitted lake and mountain scenery to become objects of mass tourism; as the twentieth century has progressed it has been air travel that opened up sunbathing on Mediterranean beaches to the workers of Northern Europe.

The post-1945 tourist boom in the Mediterranean began with a chartered Dakota and tents on a camp site in Corsica; since then numbers have increased and the accommodation improved. The increase in the number of tourists has been matched by a great growth in the number of guides, reflecting the same response to marketing opportunities that led to the publication of Wordsworth's guide. The objective of these modern tourist guides is to present local colour, usually by identifying the old and the quaint and by describing the associations of people with place. Inevitably this is frequently reduced to place detail, often little removed from a gazetteer, with a style which is flat and factual. Where, however, an author takes a broader look at a place and the people who inhabit it, some significant insights can result. Such literature will certainly have a very different appeal from the popularly orientated travel books; the writing is essentially impressionistic, aiming to capture the qualities and character of the place and its people. The key to this writing is experience, and the success of the work lies with the author's ability to allow the reader to experience the place himself. Lawrence Durrell is one such writer with this ability.

Durrell expressed this objective with the title of his book *Spirit of Place*. To him travel becomes 'a sort of science of intuitions',[16] and he sees it as his function to convey the intuitions born of experience to a wider audience. His standpoint is much the same as that adopted by regional geographers attempting to isolate the personality of an area, and he uses the same notion of a 'philosophic key' that will unlock a door to reveal the essence of a region. He believes that it is the possibility of finding this key that draws us to foreign parts, arguing that 'we travel really to try to get to grips with this mysterious quality of "Greekness" or "Spanishness"'.[17] Durrell's philosophy of the importance of place extends to his novels. He may seem to be overstating his case with his declaration that, 'Truly the intimate knowledge of landscape, if developed scientifically, could give us a political science',[18] but the strength of place in *The Alexandria Quartet* and the significance of the city and its climate not only to his characters but also to his perceptive comments on social interaction and relationships, show his belief in the power of landscape to shape people.

In terms of tourism, however, Durrell's novels take second place to his autobiographical accounts of life on the Greek islands and his travel essays which largely drew on his experience in France. Durrell had moved to Corfu in 1935 but on the fall of Greece in 1941 left for Egypt. After the war Durrell spent several years on Rhodes and Cyprus before moving to France in the mid-1950s. His experiences in these places provided the material that led to his success as a travel writer. His approach, though, is far removed from that of the conventional guidebooks, and there can be no doubt that he was writing for descendants of those 'persons of taste' to whom Wordsworth had addressed himself a century earlier. Whereas Wordsworth's 'persons' had a desire to be better informed, Durrell's success and influence stemmed from his ability to bridge a cultural gap and to convey the simplicity of a Mediterranean lifestyle as an attraction to the urbanites of industrial Europe. Those for whom the rapid development of the Spanish coastline epitomised the commercialism of the simple 'sun and sea' holiday could seek refuge in the images conveyed by Durrell. Even though *Prospero's Cell* draws upon his life on Corfu just before the Second World War,[19] it none the less allows the modern tourist to relate his own holiday experiences to Durrell's lifestyle; it allows the tourist to see parallels between Durrell's taverna conversations and his own experience of eating *al fresco*, between Durrell's trips on his boat, the *Van Norden*, and his own excursions in the company of other tourists along Corfu's coastline. These parallels, and the tourist's sharing of Durrell's experiences and his insights into the Greek character, enable him to feel less of an interloper in the close island community. Durrell's writing allows the tourist to recreate the images of rusticity in holiday environments that are otherwise highly commercial, and *Prospero's Cell* is, in many respects, an aid to that self-delusion which is an integral part of an island holiday, the experience of belonging to a community so very far removed from the industrial and urbanised ones of Northern Europe.

Durrell's earlier writings on place are full of insights into Greek character and his assessment that place and people are inseparable; his later works depend far more upon cultural and social stereotypes. None the less, he manages to give the conventional image a vitality that gives it credibility. In Durrell's view the task of the travel writer is 'to isolate the germ in the people which is expressed by their landscape',[20] and his landscapes are essentially about people. *Prospero's Cell*, his account of his life on Corfu, is peopled by his friends Zarian and Stephanides, by Karaghiosis, the epitome of the Greek character, by Count D, the recluse who likes to be visited, and by many others. It is this emphasis upon the strength of personal relationships that is possibly his most important influence upon modern tourism. He describes a society in which emotions, successes, failures and aspirations are shared. To an individual brought up on personal reliance, a quality fundamental to success in modern society, the peasant communities described by Durrell offer sharing as an antidote to the pressures of living. What is more, Durrell demonstrates the essential friendliness of these communities, and the fact that an interloper can form valuable supportive relationships.

The sum total of Durrell's influence is that he too has helped create a style in tourism. Not only has his work increased the appeal of the Greek islands, he has also introduced the exoticism of native life as an essential ingredient of the fashionable modern holiday. However, while Scott Fitzgerald's popularisation of the sunshine and sea combination was marketed in conventional terms which led to extensive hotel development along Mediterranean beaches, such a response was inimical to the values expressed by Durrell. Hotels isolate, Durrell exhorts contact; hotels internationalise, Durrell seeks distinctiveness and identity; hotels seek profit, Durrell values people. The new fashion, whose appeal Durrell has identified, analysed and popularised, is for the villa holiday and living as part of the community.

Conclusion

Literature has not generally been used by geographers as a means of analysing those economic and social events and trends that have traditionally been within their sphere of competence, but when we are concerned with an activity such as tourism which owes its whole existence to people's images – their anticipation and responses to fashion – then we can appreciate that a knowledge of literature can cast light on the reasons for stylistic appeal and for stylistic change.

This chapter has inevitably concerned itself only with tourism of a time and place that is open to the more affluent members of society since it is these who, reflecting literary influences, create the fashion and set the style. The resulting tourism is a combination of three elements – type, place and time. 'Type' can be divided crudely into scenic tourism, beach tourism and cultural or educational tourism; 'place' and 'time' are self-explanatory. We have seen that literature has played a significant role in determining fashion in each of these elements, and in so doing has played its part in determining the places to be, and the activities to do, for those who are followers of fashion. Wordsworth's verse praised the natural world, and in particular praised a landscape of lakes and mountains in such a way that it encouraged the development of scenic tourism. His significance cannot be underestimated. He wrote of everyday objects and emotions in everyday language, and thus proved to be the first effective communicator of a philosophy that provided the basis for scenic tourism generally and Lake District tourism in particular.

Scott Fitzgerald too influenced tourist trends in that he took the American idea of a beach holiday, married it to a European location and popularised a season of the year. Prior to the First World War, exposure on European beaches was still minimal. Beauty was very much a class phenomenon characterised by smooth skins and pale complexions, which contrasted with the weather-beaten features of those who worked with their hands. In the 1920s tastes changed. After the movies, sunbathing was arguably one of the first major cultural influences to cross the Atlantic from west to east. Scott Fitzgerald, in his life and in his work, reflected and made fashionable these new attitudes, and in so doing turned a winter season for the rich into a summer season for those with aspirations. In an age when transport had opened the lakes and mountains of Northern Europe to all, Scott Fitzgerald helped introduce the suntan as a mark of affluence and thus the Mediterranean as a tourist location.

Durrell's influence, like that of Scott Fitzgerald, owes much to the fact that the prevailing style had become too widely available. As a style becomes accessible to all, its value as the currency of fashion becomes debased and fashion leaders and creators seek new ways of expressing their taste and influence. When day trippers reached the Lake District, the fashionable wintered in the Mediterranean; when mass commercialism brought Spanish sun and beaches within popular grasp, Durrell's ethnic lifestyle provided the basis for a new fashion in which pleasure was sought in temporary culture shock and in which small groups replaced the anomie of modern society. The themes of this new style were simplicity and identity, and Durrell's writing lays the foundation of a tourism that in its rusticity and scale is the inevitable reaction to the massive Mediterranean beach complexes. The success of fashion, however, is that it sows the seeds of its own destruction; the tragedy of the search for simplicity is that tourism must inevitably destroy many of the communities which taste currently cultivates.

Notes

1. D. Daiches and J. Flower, *Literary Landscapes of the British Isles* (Paddington Press, London, 1979).

2. Norman Lacey, *Wordsworth's View of Nature and its Ethnical Consequences* (Archon, Hamden, Conn., 1965), p. 117.

3. 'To Joanna', lines 56–65. All quotations from Wordsworth's poems are taken from *Wordsworth Poetical Works* (ed. Thomas Hutchinson, revised by Ernest de Selincourt, Oxford University Press, London, 1975).

4. *The River Duddon* III, lines 1–4.

5. For a more detailed discussion of tourism in the Lake District from earliest times see Norman Nicholson, *The Lakers: The Adventures of the First Tourists* (Hale, London, 1972). Nicholson has a chapter (Chapter 10) on 'The Wordsworths'.

6. Graham McMaster, *Penguin Critical Anthologies: William Wordsworth* (Penguin, Harmondsworth, 1972), p. 24.

7. The 1820 version was entitled *A Topographical Description of the Country of the Lakes, in the North of England*; in 1822 and 1823 it was called *A Description of the Scenery of the Lakes in the North of England*, and in 1835 it had become *A Guide Through the District of the Lakes in the North of England*.

8. This was popularly known as *Hudson's Guide* but its full title was *A Complete Guide to the Lakes comprising Minute Directions for the Tourist, with Mr. Wordsworth's Description of the Scenery of the Country etc. and three letters on the Geology of the Lake District by the Rev. Professor Sedgwick.*

9. From the *Guide to the Lakes*, lines 4–8. All quotations from the *Guide* are taken from *The Prose Works of William Wordsworth*, Volume II (eds. W. J. B. Owen and Jane Worthington Smyser, Oxford University Press, London, 1974).

10. 'The Brothers', lines 1–10.

11. From a letter to the *Morning Post*, reprinted in Owen and Smyser, *Prose Works*, Volume III, p. 341.

12. *This Side of Paradise* and *The Beautiful and The Damned* were published in 1920 and 1922 respectively, and *Flappers and Philosophers* and *Tales of the Jazz Age* in 1921 and 1922.

13. F. Scott Fitzgerald, *Tender is the Night*, 1934 (The Bodley Head Scott Fitzgerald, Volume 2, 1964), p. 9.

14. Ibid., p. 25.

15. Ibid.

16. Lawrence Durrell, 'Landscape and Character' in *Spirit of Place* (ed. Alan G. Thomas, Faber & Faber, London, 1975), p. 160.

17. Ibid., p. 157.

18. Ibid.

19. Lawrence Durrell, *Prospero's Cell* (Faber & Faber, London, 1945).

20. Durrell, *Spirit of Place*, p. 156.

9 JOHN STEINBECK'S *THE GRAPES OF WRATH* AS A PRIMER FOR CULTURAL GEOGRAPHY

Christopher L. Salter

There is no need to write additional textbooks in cultural geography. All the messages of the profession are already committed to ink. The motivations, processes, patterns and the consequences of human interaction with the landscape have all been discovered and chronicled with grace and clarity. Authors dedicated to the comprehension and elucidation of order within the overtly haphazard flow of human events have given academics the materials needed to profess the patterns which illustrate this order. We fail, however, as scholars to make adequate use of these data for the simple reason that this material is labelled 'fiction'.[1]

Fiction in its primary meaning denotes invention. Ironically, the process of invention in the human species is one of the most consistently lauded acts that we can be associated with. Invention in professional fields is celebrated as creativity and insight. The same act in the commercial world generates considerable cash. And even greater commendation is heaped on the inventor if the product of his or her imagination can be used in fields other than the inventor's own. That becomes the product of genius.

Yet, invention in the field of creative writing – the fiction of imaginative observation – is too often held to be non-transferable. It is in this domain that – for academics – fiction can assume its pejorative connotation. Constrained by a narrowly conceived framework of objectivity, the inventions of a good literary mind may indeed be unsuitable source material for research or teaching. The nature of human experience, however, whether that between fellow humans or in their relationship with place, cannot be captured in a rigidly objective framework. Imaginative literature articulates the kaleidoscope of human experience and the teacher of cultural geography who adopts an alternative artistic or humanistic stance to such material may himself be deemed inventive. The arguments in favour of such a literary teaching programme for cultural geography may be briefly enumerated.

Consider first the learning atmosphere engendered by the use of a novel in addition to, or in lieu of, an orthodox text in cultural geography. Largely because a novel is a work of fiction, a reader slips

into the narrative with a curious mind. The interest in understanding
the author's work derives from an informal competition between the
reader and the author. What is the message here? asks the student.
Can the author make me concerned enough about it to give of my mind,
wit and time?

In the same situation, a traditional textbook would be anticipated as
a collection of facts strung together as beads on a time line. The
ambition of the reader would be most probably focused upon retention
rather than upon understanding, hence diminishing creative speculation
regarding the implications of the material presented. A novel gains
strength because it may fire one's imagination through subtle allusion
and illusion. A textbook, on the other hand, damps down the same fire
through demands for inclusion and conclusion.

Literary fiction works well with cultural geography because the
substance of both endeavours is life itself. The capacity for attitudes
that shape environmental manipulation are present in all people,
whether their perspectives emerge from an author's pen or a social
scientist's interview data. The task of the cultural geographer is the same
regardless of the data base: to mould individual specifics into understandable, reliable predictabilities. If the target audience of such an
intellectual effort becomes involved in and concerned with the specifics
of the human process, it will be easier for the teacher to instruct in the
larger realities. These mundane specifics are exactly the material the
novelist employs to create his fiction.

The Grapes of Wrath as Example

Because so many works of fiction are fundamentally genuine in their
description of people and place, the range of works available to a class
is large. In the selection of a novel for such an experiment, therefore,
the teacher's best guide is his or her own personal preference in authors
and settings.[2] The re-reading of a classic, with an eye focused upon the
themes of cultural geography, may well produce new insights and new
analysis.

For this chapter, I have chosen to use the major work by the
American author John Steinbeck. In his 1939 *The Grapes of Wrath*[3]
several significant criteria are met. In the first place, the novel is the
product of vital and personal fieldwork by the author. He was writing
of a world that he knew very intimately.[4] Secondly, he has his
characters move through a variety of distinct physical and cultural

regions. Such movement adds variety to the geographic observations that are potential in a novel. In the third place, Steinbeck employs a useful literary convention in the book in his use of inner chapters. These short episodes that break up the specific narrative of the Joad family provide the reader with an overview of the cultural landscape during the time the fiction of the novel takes place. By interrupting the flow of the personal narrative of the primary family, Steinbeck effectively causes the reader to back away from the specific incidents of his main family and, instead, to view the difficulties of these people as part of a larger social fabric. While such a technique is far from unique to Steinbeck, the structure of *The Grapes of Wrath* is particularly well suited to developing societal themes as opposed to simply personal themes.

The final reason for the selection of *The Grapes of Wrath* is that if members of a class become interested in Steinbeck's style and concerns, there exists a large corpus of work that may be subsequently turned to. Once a student begins to read fiction with part of his mind searching out distinctive landscapes and culture systems, education has essentially reached a new and higher level. It is fitting that a cultural geographer should play some role in that attainment.

In *The Grapes of Wrath* three major theme areas in cultural geography are particularly evident for elaboration. Human mobility — the kinetic energy of so much landscape transformation — plays a dominant role in Steinbeck's narrative. Tensions between competing modes of land use spark much of the drama in the novel, while illustrating the social consequences of such variant decisions in economic patterns. Finally, the specific social and spatial configurations encountered by the Joads during their epic move from Oklahoma to California serve as vignettes around which a geographer can structure expanded explanations of human transformation of the land. In this paper, our concern is focused upon the themes of human mobility.

Human Mobility as a Dominant Force in Cultural Geography and in *The Grapes of Wrath*

Systems of belief as well as spatial order are constantly subject to change. The intrusion of competing systems — through the processes of migration and expanded communications — is a key force in such change. To the cultural geographer human mobility affords a thematic domain that embraces environmental perception, regional, cultural and economic variation, problem landscapes that impede migration and mobility, as well as contesting systems of custom, government and settlement. Steinbeck weaves all of these elements into his novel of America in the 1930s. By sorting out several of the most significant sub-themes in this universe of movement, an orderly analysis of both the novel and this aspect of cultural geography is possible.

Information Fields and the Selection of a Goal Area

Axial to any model of decision-making in a migration scenario is the process of deciding where one is to go when the hearth area no longer accommodates a person or a people. The information available in that selection of a goal area derives from hearsay, media messages, feedback from earlier migrants, reading and current folklore. This initial consideration of the choice of goal area is appropriate in our analysis of *The Grapes of Wrath* because so great a part of the novel is concerned with the search for a haven. In this process, the Joads and their fellow travellers were continually confronted with the riddle of 'where to go'. The image of California that was known in Oklahoma was one of a tarnished Eden. Although news of the fields, the fruit, the opportunity for land and the generalised abundance of everything in California was widespread, there were increasing rumours of the existence of an ugly dark side to this vision. In the two passages below, members of the Joad family touch on the emotional extremes of the images that came to the Okies as they were forced to pack up their few belongings and set out for new lands. Grampa speaks first, representing the popular image of an abundant reward awaiting those who chose California for a goal area in the departure from the drought-stricken Great Plains.

> The old man thrust out his bristly chin, and he regarded Ma with his shrewd, mean, merry eyes. 'Well, sir,' he said, 'we'll be a-startin' 'fore long now. An', by God, they's grapes out there, just a-hangin' over inta the road. Know what I'm a-gonna do? I'm gonna pick me a wash tub full a grapes, an' I'm gonna set in 'em, an' scrooge aroun', an' let the juice run down my pants.'[5]

Just before that passage, Ma Joad allowed herself to speculate on some of those same images when she allowed herself to think of

> how nice it's gonna be, maybe, in California. Never cold. An' fruit ever'place, an' people just bein' in the nicest places, little white

houses in among the orange trees. . . An' the little fellas go out an' pick oranges right off the tree. They ain't gonna be able to stand it, they'll get to yellin' so.[6]

But a parallel ambivalence the family felt about California and the decision to uproot and head west is evident in this turn of the conversation between Ma and her son, Tom, who has just been paroled from prison.

'I knowed a fella from California. He didn't talk like us. You'd of knowed he come from some far-off place jus' the way he talked. But he says they's too many folks lookin' for work right there now. An' he says the folks that pick the fruit live in dirty ol' camps an' don't hardly get enough to eat. He says wages is low an' hard to get any.'
A shadow crossed her face. 'Oh, that ain't so,' she said. 'Your father got a han'bill on yella paper, tellin' how they need folks to work. They wouldn't go to that trouble if they wasn't plenty work. Costs 'em good money to get them han'bills out. What'd they want ta lie for, an' costin' 'em money to lie?'[7]

This refusal of Ma to accept the information that Tom brought from prison — from one who had witnessed a different image of the state — is significant, for it is illustrative of the manner in which the migrant begins to exclude information that threatens the positive image of the goal area. Once plans for a migration have been made, blinkers are put on in an attempt to disallow any dilution of that resolution. As one ponders the trauma of an uprooting from the home and place that has served as home for decades for a family, such a response becomes increasingly understandable. The psychological costs of such a move provide the second theme for consideration.

The Psychological Costs of Mobility

The whole tone of movement in *The Grapes of Wrath* is one of regretful departure. Except for occasional, if powerful, allusions to a better life in California voiced by the young Joads and Grampa, the family leaves their Oklahoma sharecropping past with reluctance.[8] This wrenching of the move is not limited to the Joads. Steinbeck creates one of his strongest characters in the person of Muley, one of the few people the reader meets who has decided not to move to California. The combination of Ma Joad and Muley — although they never share the same stage — is effective in making an observer realise that this process of uprooting at a time of crisis has deep and profound psychological costs far beyond the economic dislocation associated with such a migration.

In the one case — that of Ma Joad — there is a buffer created by the fact that she is taking her family with her. She will at least be able to maintain basic associations with the people who are most important in her life. In the case of Muley, Steinbeck creates a character who is cast adrift entirely. In his attempts to maintain his attachment to the past and his personal tradition, he reverts to a near-primitive, taking solace in his attempts to make trouble for the agents of change sent by the banks and the police.[9]

This trauma of movement derives also from the sheer financial burdens of relocation. For the Okies, to whom Steinbeck gives his novel, there were few resources available to lighten the burden of the move. Human resources were the greatest riches these farmers had, and it was these very family members who found themselves at odds with each other because of decisions about where to go or how to travel or what to drive. Details in the inner chapters as well as the Joad narrative point out to the reader the complexity of human mobility, a complexity that involves both the emotions and the finances of individuals and families. Such an understanding helps an observer to realise the gravity of the decision to uproot and move on.

Passages that demonstrate the power of these themes of dislocation come from conversations between Tom and Muley, Ma and Tom, and descriptions of the car lots in the inner chapters. In the first excerpt below, Muley, Tom and Casy are talking around a small fire on the nearly deserted farmlands of the town where Tom and Muley were raised. Muley is getting increasingly excited as he explains how he felt as he saw his house emptied, his parents leave and his farmland all reduced to a sameness under the power of the new tractors on the land.

'I wanta talk. I ain't talked to nobody. If I'm touched, I'm touched, an' that's the end of it. Like a ol' graveyard ghos' goin' to neighbors' houses in the night. Peters', Jacobs', Rance's, Joad's; an' the houses all dark, standin' like miser'ble ratty boxes, but they was good parties an' dancin'. An' there was meetin's and shoutin' glory. They was weddin's, all in them houses. An' then I'd want to go in town an' kill folks. 'Cause what'd they take when they tractored the folks off the lan'? What'd they get so their 'margin a profit' was safe? They got Pa dyin' on the groun', and Joe yellin' his first breath, an' me jerkin' like a billy goat under a bush in the night. What'd they

get? God knows the lan' ain't no good. Nobody been able to make a
crop for years. But them sons-a-bitches at their desks, they jus'
chopped folks in two for their margin a profit. They jus' cut 'em in
two. Place where folks live is them folks. They ain't whole, out
lonely on the road in a piled-up car. They ain't alive no more. Them
sons-a-bitches killed 'em.' And he was silent, his thin lips still
moving, his chest still panting. He sat and looked down at his hands
in the firelight. 'I-I ain't talked to nobody for a long time,' he
apologized softly. 'I been sneakin' around like a ol' graveyard
ghos'.'[10]

Ma shows some of the same unprecedented irritation as she attempts
to organise her family's goods for the uncertain trip west.

'Ma,' he said, 'you never was like this before!' Her face hardened
and her eyes grew cold. 'I never had my house pushed over,' she
said. 'I never had my fambly stuck out on the road. I never had to
sell-ever'thing. . .'[11]

A little later on, Ma looks at the pile of household goods that the
family cannot carry and asks a most critical question: 'How can we live
without our lives? How will we know it's us without our past? No.
Leave it. Burn it.'[12]

The economic difficulties of the sharecroppers' moves are compounded by the size of the multitude that is taking to the road. In one
of the novel's most effective inner chapters, Steinbeck gives the reader
a close-up of a used car lot. In the excerpt below, the reader gains a
strong sense of how victimised migrants can be when some environmental or social catastrophe forces so many people to move
simultaneously that individuals lose any economic leverage they might
otherwise have.

What you want is transportation, ain't it? No baloney for you. Sure
the upholstery is shot. Seat cushions ain't turning no wheels over.
　　Cars lined up, noses forward, rusty noses, flat tires. Parked close
together.
　　Like to get in to see that one? Sure, no trouble. I'll pull her out
of the line.
　　Get 'em under obligation. Make 'em take up your time. Don't
let 'em forget they're takin' your time. People are nice, mostly. They
hate to put you out. Make 'em put you out, and then sock it to 'em.

Cars lined up, Model T's, high and snotty, creaking wheel, worn bands. Buicks, Nahes, De Sotos.

Yes, sir. '22 Dodge. Best goddam car Dodge ever made. Never wear out. Low compression. High compression got lots a sap for a while, but the metal ain't made that'll hold it for long. Plymouths, Rocknes, Stars.[13]

This reality is further exploited in scenes from town when the Joads attempt to sell their household goods. There is little return on a lifetime of accumulation when the entire farm community is trying to sell its untransportable furniture and farm implements at the same time.[14]

Notwithstanding the agonies associated with the move, the people do uproot and leave. Not only is that departure essential to the novel, but it is a reality in consideration of human mobility. Even in the face of the hardships narrated in *The Grapes of Wrath*, as well as in the myriad others that exist in all migration, people do carry through with the plan to try some other place, hoping that the change will bring them more benefit than cost. In that movement, the people and the landscape are changed. Such change introduces a third major theme area in the study of human mobility from the perspective of cultural geography.

Personal and Landscape Changes Associated with the Process of Human Mobility

It is not the migrants alone who are changed by the process of migration. Demographic shifts modify the complexion of the hearth area as well as the goal area. If the process or the novel that we are studying is based on an individual's experience, then the scale of such impact is probably small. However, as the size of the population in motion grows, the magnitude of the influence created by the migration increases accordingly. Additionally, the corridor through which the movement takes place is bombarded with new demands, demands which are seldom met with welcome.

The nature of this intermediate response is captured in a number of episodes by Steinbeck, but perhaps most powerfully in his scenes from a Highway 66 diner. The excerpt below shows the pensive nature of a waitress, for example, who has had her entire universe confused by all of the people and families moving west.

Flies struck the screen with little bumps and droned away. The compressor chugged for a time and then stopped. On 66 the traffic whizzed by, trucks and fine stream-lined cars and jalopies; and they went by with a vicious whiz. Mae took down the plates and scraped the pie crusts into a bucket. She found her damp cloth and wiped the counter with circular sweeps. And her eyes were on the highway, where life whizzed by.[15]

This life that whizzed by in Mae's eyes brought changes all along Highway 66. Traditionally free services such as water and air at the roadside gas stations became so overused that dealers began to charge for such facilities. Used tires, fan belts and engine service all became more expensive as service-station owners' reactions went from irritation and suspicion to entrepreneurial opportunism, creating an image of their services through roadside junk.

The truck drove to the service-station belt, and there on the right-hand side of the road was a wrecking yard — an acre lot surrounded by a high barbed-wire fence, a corrugated iron shed in front with used tires piled up by the doors, and price-marked. Behind the shed there was a little shack built of scrap, scrap lumber and pieces of tin. The windows were windshields built into the walls. In the grassy lot the wrecks lay, cars with twisted, stove-in noses, wounded cars lying on their sides with the wheels gone. Engines rusting on the ground and against the shed. A great pile of junk; fenders and truck sides, wheels and axles; over the whole lot a spirit of decay, of mold and rust; twisted iron, half-gutted engines, a mass of derelicts.[16]

The images of the migration process — whether in the novel or in a teacher's reality — are frequently tied to this very problem of transportation. Just as the covered wagon of a century earlier became metonymy for the entire process of American westward expansion and settlement, the jalopy with its children, inverted chairs, wooden barrels of dishes, pans and rags became the visual signature of the migration depicted by Steinbeck in *The Grapes of Wrath*. These rigs required impressive levels of self-reliance and inventiveness, qualities that begin to emerge in any migration as the movers are forced to deal with the landscapes and the people they encounter in their flight from the past.

Thus they changed their social life — changed as in the whole universe only man can change. They were not farm men any more, but migrant men. And the thought, the planning, the long staring silence that had gone out to the field, went now to the roads, to the distance, to the West. That man whose mind had been bound with acres lived with narrow concrete miles. And his thought and his worry were not any more with rainfall, with wind and dust, with the thrust of the crops. Eyes watched the tires, ears listened to the clattering motors, and minds struggled with oil, with gasoline, with the thinning rubber between air and road. Then a broken gear was tragedy. Then water in the evening was the yearning, and food over the fire. Then health to go on was the need, and strength to go on, and spirit to go on. The wills thrust westward ahead of them, and fears that had once apprehended drought or flood now lingered with anything that might stop the westward crawling.

The camp became fixed — each a short day's journey from the last.[17]

To accompany this psychological change in attitude came an associated change in setting. Finding a few physical elements deemed necessary for a nighttime haven — water, a little firewood and perhaps a nearby dump for scavenging — these people began a pattern of creation anew each night. As some of these 'Hoovervilles' became established, they began to take on a geography of their own. Such settlements are a *bona fide* segment of the landscape of change in this human movement.

There was no order in the camp; little gray tents, shacks, cars were scattered about at random. The first house was nondescript. The south wall was made of three sheets of rusty corrugated iron, the east wall a square of moldy carpet tacked between two boards, the north wall a strip of roofing paper and a strip of tattered canvas, and the west wall six pieces of gunny sacking. Over the square frame, on untrimmed willow limbs, grass had been piled, not thatched, but heaped up in a low mound. The entrance, on the gunny-sack side, was cluttered with equipment. A five-gallon kerosene can served for a stove. It was laid on its side, with a section of rusty stovepipe thrust in one end. A wash boiler rested on its side against the wall; and a collection of boxes lay about, boxes to sit on, to eat on. A Model T Ford sedan and a two-wheel trailer were parked beside the shack, and about the camp there hung a slovenly despair.

Next to the shack there was a little tent, gray with weathering, but neatly, properly set up; and the boxes in front of it were placed against the tent wall. A stovepipe stuck out of the door flap, and the dirt in front of the tent had been swept and sprinkled. A bucketful of soaking clothes stood on a box. The camp was neat and sturdy. A Model A roadster and a little home-made bed trailer stood beside the tent.

And next there was a huge tent, ragged, torn in strips and the tears mended with pieces of wire. The flaps were up, and inside four wide mattresses lay on the ground. A clothes line strung along the side bore pink cotton dresses and several pairs of overalls. There were forty tents and shacks, and beside each habitation some kind of automobile. Far down the line a few children stood and stared at the newly arrived truck, and they moved toward it, little boys in overalls and bare feet, their hair gray with dust.[18]

Every choice humankind makes for the manipulation of an environment sets in motion a tension. The prior state of development at any given point was either natural (increasingly unlikely) or represented some other human wish or design. Cultural geography finds much of its meaning from the process of analysis and evaluation of such a change process, and resultant environments. Questions of natural conditions, technology, economic systems, social attitudes, demography, custom and, finally, special forces of the moment all must be factored into any equation that attempts to explain change. The creation of the Hoovervilles — noted above — was one of the most explicit signals to the Californians that their state was destined to undergo marked social and spatial change in response to this migration stream that had been initiated more than a thousand miles away. These response patterns in the goal area introduce us to our final theme in the cultural analysis of mobility in the novel.

Cultural Response Patterns in the Goal Area

Although public response to the 1939 publication of *The Grapes of Wrath* was vastly supportive in terms of book purchases, Steinbeck found himself very uncomfortable in California.[19] His portrayal of his state's citizens as being unsympathetic, avaricious, even malicious towards this stream of migrants from Oklahoma and other states of the Dust Bowl, sorely wounded the pride of the folk with whom Steinbeck had grown up.[20] Criticism was focused upon his inability to acknowledge the impact this immigration of penniless, rural and distraught folk would have on the social order of California. The Great Depression, although not damaging this western state as profoundly as it had other parts of the country, had already taxed municipal and state agencies to the margin of their abilities to cope with unemployed, angry people.

The spectre of additional hundreds of thousands of like souls, but souls
new to the state, making similar demand on modest resources excited
no small anxiety in the eyes of nearly all Californians.

Steinbeck establishes the potentially explosive mood of this tension
between the migrants and Californian patterns of farming in one of his
strongest inner chapters. He describes the capital-intensive nature of
local farming, pointing out the fixed expenses for chemical fertilisers,
spraying and irrigation technology. The depressed prices, however, of
the late 1930s drove prices below a level that generated essential
income for the strictly managed farms. Instead of selling at such a level,
owners decided to destroy fruit in order

> to keep up the price, and this is the saddest, bitterest thing of all.
> Carloads of oranges dumped on the ground. The people came from
> miles to take the fruit, but this could not be. How would they buy
> oranges at twenty cents a dozen if they could drive out and pick
> them up? And men with hoses squirt kerosene on the oranges, and
> they are angry at the crime, angry at the people who have come to
> take the fruit. A million people hungry, needing the fruit — and
> kerosene sprayed over the golden mountains. . .
>   There is a crime here that goes beyond denunciation. There is a
> sorry here that weeping cannot symbolize. There is a failure here
> that topples all our success. The fertile earth, the straight tree rows,
> the sturdy trunks, and the ripe fruit. And children dying of pellagra
> must die because a profit cannot be taken from an orange. And
> coroners must fill in the certificates — died of malnutrition —
> because the food must rot, must be forced to rot. . . In the souls of
> the people the grapes of wrath are filling and growing heavy, growing
> heavy for the vintage.[21]

One of the classic confrontations of the new migrants with the old
order in California came in the hiring of pickers for the ripening fruits
up and down the central valley of the state. Contractors would go to
the Hoovervilles, offer work at 30 cents an hour at some distant farm
and give families directions on how to get there. Arriving at the place,
having exhausted resources and given up space at the camp, people
would be told that the rate was only 15 cents an hour — and there
would be a surplus of workers even at that rate. Occasionally,
migrants who had been caught up in this painful misrepresentation
several times would attempt to dissuade the Okies from falling into the
trap. The scene below illustrates one such happening.

'You men want to work?' . . . men from all over the camp moved near. One of the squatting men spoke at last. 'Sure we wanta work. Where's at's work?'

'Tulare County. Fruit's opening up. Need a lot of pickers.'

Floyd spoke up. 'You doin' the hiring?'

'Well, I'm contracting the land.'

The men were in a compact group now. An overalled man took off his black hat and combed back his long black hair with his fingers. 'What you payin'?' he asked.

'Well, can't tell exactly, yet. 'Bout thirty cents, I guess.'

'Why can't you tell? You took the contract, didn't you?'

'That's true,' the khaki man said. 'But it's keyed to the price. Might be a little more, might be a little less.'

Floyd stepped out ahead. He said quietly, 'I'll go, mister. You're a contractor, an' you got a license. You jus' show your license, an' then you give us an order to go to work, an' where, an' when, an' how much we'll get, an' you sign that, an' we'll all go.'

The contractor turned, scowling. 'You telling me how to run my own business?'

Floyd said, ''F we're workin' for you, it's our business too. . .'

Floyd turned to the crowd of men. They were standing up now, looking quietly from one speaker to the other. Floyd said, 'Twicet now I've fell for that. Maybe he needs a thousan' men. He'll get five thousan' there, an' he'll pay fifteen cents an hour. An' you poor bastards'll have to take it 'cause you'll be hungry. 'F he wants to hire men, let him hire 'em and write out an' say what he's gonna pay. Ast ta see his license. He ain't allowed to contract men without a license.'[22]

Clandestine farming was another point of contention between the uprooted farmers of the Dust Bowl and the California residents who became increasingly uneasy about the threat to their patterns of agriculture.

Now and then a man tried; crept on the land and cleared a piece, trying like a thief to steal a little richness from the earth. Secret gardens hidden in the weeds. A package of carrot seeds and a few turnips. Planted potato skins, crept out in the evening secretly to hoe in the stolen earth.

. . . Secret gardening in the evenings, and water carried in a rusty can.

And then one day a deputy sheriff: Well, what you think you're doing?

I ain't doin' no harm.

I had my eye on you. This ain't your land. You're trespassing.

The land ain't plowed, an' I ain't hurtin' it none.

You goddamned squatters. Pretty soon you'd think you owned it. . . Get off now. And the little green carrot tops were kicked off and the turnip greens trampled. . . Did ya see his face when we kicked them turnips out? Why, he'd kill a fella soon's he'd look at him. We got to keep these here people down or they'll take the country . . . Outlanders, foreigners . . . Sure, they talk the same language, but they ain't the same. Look how they live. Think any of us folks'd live like that? Hell, no![23]

In *The Grapes of Wrath* — and even more so in the reality of the migration — there were instances of a more welcoming response to the migrants. People were able to see these families as fundamentally hardworking farm people who had been set in motion by the extraordinary combination of the natural forces of the drought and dust conditions of the Great Plains and the economic chaos of the Depression. Steinbeck portrays one such sympathetic farmer in the person of a Mr Thomas who hires men from the government camp of Weedpatch. He is a small farmer, deeply dependent upon a credit line from his bank, even though he appears to have been an efficient farmer. In this excerpt Mr Thomas has just told his small work crew that the 30 cents an hour they had been getting was being reduced to 25 cents that morning.

Timothy said, 'We've give you good work. You said so yourself.'

'I know it. But it seems like I ain't hiring my own men any more.' He swallowed. 'Look,' he said, 'I got sixty-five acres here. Did you ever hear of the Farmers' Association?'

'Why, sure.'

'Well, I belong to it. We had a meeting last night. Now, do you know who runs the Farmers' Association? I'll tell you. The Bank of the West. That bank owns most of this valley, and it's got paper on everything it don't own. So last night the member from the bank told me, he said, "You're paying thirty cents an hour. You'd better cut it down to twenty-five." I said, "I've got good men. They're worth thirty." And he says, "It isn't that," he says. "The wage is twenty-five now. If you pay thirty, it'll only cause unrest. And by the way," he says, "You going to need the usual amount for a crop loan next year?"' Thomas stopped. His breath was panting through his lips. 'You see? The rate is twenty-five cents — and like it.'[24]

Conditions, then, at the place of arrival paralleled the set of conditions that drove the farmers off the land and onto Highway 66. The banks that had been seeking the 'margin of profit' that Muley spoke of in Oklahoma appeared to be in charge of farming decisions on small and large landholders alike in California. The vagaries of nature that brought drought and turned sharecropping counter-productive also produced the prolonged rains and floods that end the book with the remnants of the Joad family stranded in a desolate boxcar, still out of touch with the land that they had set out for. Although migration had been undertaken, the same paucity of options that faced these families at the outset seems to characterise their future as the novel is closed.

That, perhaps, is the fiction that cultural geographers should do battle with in their instruction on human mobility. Even with the specifics broadly varying from case to case, the fact of migration does open new options. People change in the process of movement; places change as the migrants grow more familiar with the setting and cultural fabric of these new locales. In a harsher sense, the least adaptive of the initial migrants have probably left the migration stream, diminishing the competition in the search for support at the final destination.

New skills are learned by the migrant farmers as they leave the land to their past and find outlets for their ambition in the cities. New settlement features grow up around the migrants who finally do create a marginal haven for themselves and their families, and bring their music, foods, clothes and language into the society of the new setting. Almost nothing is able to escape some modification in the face of a migration stream as robust and intense as this particular American flight from the Great Plains in the mid-1930s. The event in itself is a dramatic exercise in the elements of cultural geography.

Conclusion

To the cultural geographer, then, lessons from the landscape and human movement in *The Grapes of Wrath* provide focus for instruction in migration, settlement forms, economic systems, cultural dualism, agricultural land use patterns, transportation technology and social change. To the reader of creative fiction, these same realities generally lie scattered within the pages of this epic of one family's unsuccessful search for a new beginning. But, to the reader of fiction who is also attempting to comprehend something of the underlying systems in this chaos of conflict and flight, the study of this novel provides a window on geographic phenomena broadly ranging from mental maps to economic infrastructures.

In the face of such complexity, effective thinking – let alone instructing – calls for the use of all the human resources available. Evocative fiction in creative literature is one of these resources. Such work, when read with a searching mind, and then re-read with a disciplined perspective, is capable of illustrating patterns, preferences and problems of humankind. And it conveys all of these dynamics with vitality. Cultural geographers – in their ambitious quest for the understanding of human society and cultural landscapes – would do well to capture and utilise such dynamics and such vitality. John Steinbeck's *The Grapes of Wrath* is one volume that possesses both qualities in such abundance that it serves provocatively as a primer for cultural geography.

Notes

1. There is a broad literature discussing the use of fiction in both teaching and research. Some of the items of interest in the literature of fiction include A. J. Lamme III, 'The Use of Novels in Geography Classrooms', *Journal of Geography*, vol. 76, no. 2 (February 1977), pp. 66–8; D. W. Meinig, 'Environmental Appreciation: Localities As a Humane Art', *Western Humanities Review*, vol. 25 (Winter 1971), pp. 1–11; C. L. Salter and W. J. Lloyd, 'Landscape in Literature', *Resource Papers for College Geography* (Association of American Geographers, Washington DC), no. 76–3 (1977); Sherman E. Silverman, 'The Use of Novels in Teaching Cultural Geography of The United States', *Journal of Geography*, vol. 76, mp/4 (April/May 1977), pp. 140–6; C. L. Salter, 'Signatures and Settings: One Approach to Landscape in Literature' in Karl W. Butzer (ed.), *Dimensions of Human Geography* (University of Chicago Press, Chicago, 1978), pp. 69–83; John Conron, *The American Landscape* (Oxford University Press, Chicago, 1973).

2. Salter and Lloyd, 'Landscape in Literature', pp. 29–30, includes a number of useful references for searching out fiction to coincide with region and, sometimes, theme of a teacher or researcher.

3. John Steinbeck, *The Grapes of Wrath* (Viking Press, New York, 1939).

4. Elaine Steinbeck and Robert Wallsten (eds.), *Steinbeck: A Life in Letters* (Penguin, New York, 1976). This collection has two sections that deal with the period of preparation for *The Grapes of Wrath* (pp. 57–190). These Steinbeck letters and their associated discussion provide the reader with a strong sense of how immediate the Oklahoma sharecropper migration was to Steinbeck during the 1930s. See also John Steinbeck, *Their Blood is Strong* (Simon J. Lubin Society, San Francisco, 1938); Peter Lisca, 'The Grapes of Wrath' (pp. 75–101), and George Bluestone, 'The Grapes of Wrath' (pp. 102–21) in Robert Murray Davis (ed.), *Steinbeck: A Collection of Critical Essays* (Prentice-Hall, Englewood Cliffs, NJ, 1972).

5. Steinbeck, *Grapes*, p. 126.

6. Ibid., p. 124.

7. Ibid.

8. Grampa fills the role of a particularly tragic character in the early part of the novel because of the naivety of his vision of California, as well as his difficulty in comprehending the magnitude of the family move. The Joad children and their friends become caught up in the adventure of moving and exploration, also failing to sense the impending rupture in the family's situation. See Chapter 10 (pp. 122–56).

9. In Chapter 6 (pp. 54–82) Muley shows Tom and Casy how thorough his adaptation to his new life has been. He anticipates the arrival and search behaviour of the sheriff; produces rabbits that he has caught with a fierce efficiency, and he rants on and on about the bank and its determination to ruin the sharecroppers of the region for its 'margin a profit'. Steinbeck uses this character to illustrate the consequences of making a decision *not* to move away from this drought-plagued region.

10. Steinbeck, *Grapes*, pp. 70–1.

11. Ibid., p. 104.

12. Ibid., p. 120.

13. Ibid., p. 84.

14. Chapter 9 (pp. 117–21) is a short inner chapter that dramatises the deep frustration that the sharecroppers felt as they tried to sell the goods that they could not transport west. The intensity of this sorrow and anger was shown as families piled household goods in their front yards, set fire to them, and watched them burn as they climbed into their overloaded trucks and headed for their uncertain future. Peter Lisca discusses these inner chapters and their accuracy, while pointing out that Steinbeck's novel launched numerous volumes supporting and disputing the images created by his characterisation of the migration. Lisca, 'The Grapes', pp. 78–93.

15. Steinbeck, *Grapes*, pp. 220.

16. Ibid., pp. 241–2.

17. Ibid., pp. 267–8.

18. Ibid., pp. 328–9.

19. The Viking Press issued the first edition of *Grapes* in April 1939 and the book went through ten printings before the end of the year.

20. Steinbeck once revealed that an undersheriff of Santa Clara County in California – a prime agricultural county at the time – warned him to be careful because local people had plans to set up a fake rape case in order to discredit him (Steinbeck and Wallsten, *Letters*, p. 187). In a conversation with a librarian in Salinas, California, in the summer of 1979, I was told that only 'in the last few years have the townspeople taken any pride at all that John Steinbeck was born here. Before that he was seen as a disgrace.'

21. Steinbeck, *Grapes*, pp. 476–7.

22. Ibid., pp. 358–9.

23. Ibid., pp. 321–2.

24. Ibid., p. 402.

10 A SOCIAL-LITERARY GEOGRAPHY OF LATE-NINETEENTH-CENTURY BOSTON

William J. Lloyd

The closing decades of the nineteenth century witnessed the rise of the urban novel as an important form of literary expression in America. Stories focusing on the everyday aspects of life in the rapidly expanding cities were part of a movement toward greater realism in American fiction, led to a considerable extent by the novels of William Dean Howells. The city of Boston, with its firmly established literary reputation, was a particular beneficiary of this increased interest in the modern city. A substantial number of works published during the 1880s and 1890s by many different resident authors sought to describe and interpret the life and landscapes of that city.[1] Together they provide us with a literary view into the social geography of the city that is rich in the attitudes and beliefs of the times, and that helps illuminate an important dimension of the foundation of modern urban life and thought.[2] These fictional accounts are particularly intriguing in that they possess the persuasive power of creative literature to evoke strong images in the reader's mind. This capability of fiction to fashion a world possessing a stark semblance of reality was not lost on contemporary observers of the Boston scene. The relationship between the Boston that appeared in fiction and the actual city was a popular theme of books and articles, leading one civic spokesman to observe that 'fiction joins hands with actuality to build the Boston that men know'.[3]

Urban fiction represents only a small part of the body of literature that was appearing in response to the growth of the modern city, but it is especially valuable to us because it helps illuminate the everyday life of the middle class.[4] Social status and the struggle for social mobility were the key forces which bound this middle-class world together, and it was values such as these which received considerable attention from the novelists. The residential landscape often occupied a prominent part in their stories since the worth of an individual and his family was measured to a great extent by the social standing of the neighbourhood where they resided. Novelists recognised that this form of social organisation placed a premium on the ability to secure the most prestigious residence that one could afford, leading them to emphasise

Figure 10.1: A Literary-social Geography of Late-nineteenth-century Boston

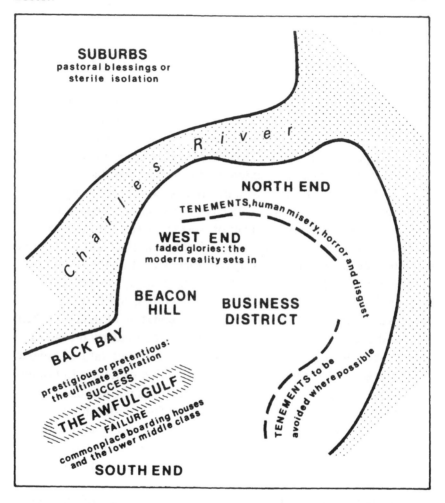

the geographical process of residential mobility and the underlying set of residential landscape images as symbols of their characters' social aspirations and standing (see Figure 10.1). The great value associated with residing in the elite Back Bay district was a common theme, so much so that this area came to serve as a standard of excellence against which to evaluate the once-fashionable West End and the unfashionable middle-class South End. Further down the social ladder, the tenements and lodging-house districts received only occasional attention within the overwhelmingly middle-class perspective of the Boston novelists. Further out towards the periphery of the metropolis, the rapidly growing suburban communities were only slowly making their significance felt on writers who were still much more at home with the city of Boston and its rich assemblage of literary associations and traditions. This composite social landscape represented a radical departure from the simpler forms of an earlier Boston, and it was viewed with some reserve by the rather conservative authors of Boston fiction. What they have left us is a vivid portrait of the middle classes trying to establish their place in a city they had created, yet one they clearly did not fully understand. Here, perhaps, we can find some evidence of the origins of that middle-class ambivalence toward the city which has occupied so much of the urban debate in the twentieth century.

The Social-literary Districts

The Prestigious Back Bay

Although the filled lands of the Back Bay district were but a couple of decades old by the late nineteenth century, Boston novelists widely recognised the area as the most prestigious residential district in the city. Authors generally praised the Back Bay and its landscape of ornate row houses and stately apartment buildings, employing such phrases as 'select regions of Back Bay', 'fashionable quarter' or 'neighbourhood of opulence'.[5] The literature left little doubt that the Back Bay had been widely adopted by wealthy Boston residents as their own distinctive district. The Back Bay place name became a symbol for high status and success in Boston, while the tangible reality of the Back Bay landscape provided a vivid sense of identity for those who resided there, just as it presented an almost insurmountable barrier to the remainder of the Boston population who were not so fortunate as to live there.

This symbolic importance of the Back Bay encouraged its use in fiction as a landscape marker that represented the ultimate goal of upwardly mobile people within Boston's middle class. The theme of intra-urban migration in the direction of the Back Bay was frequently employed by authors to illustrate their character's social ambitions, the most famous example being Howells' *The Rise of Silas Lapham*, a story of a family that was newly rich but otherwise lacking in the qualities that were worthy of a high social status.[6] The value that people like the Laphams placed on Back Bay residence seemed to grow in direct proportion to the difficulties they faced in getting there. Howells suggested that people such as the Laphams were prevented by the circumstances of their rapid rise to wealth from ever fully appreciating

the true nature of the social goals they were pursuing. This critical
shortcoming was symbolised by the lavish new house they were having
constructed in anticipation of their move from the South End to the
Back Bay, for, as Howells noted, 'they had crude taste in architecture
and they admired the worst'.[7] Seemingly predestined for failure,
Lapham accidentally set fire to his new home on the eve of its
completion and thus effectively put an end to both his wealth and his
dream of social acceptance in the Back Bay. Elusive as the goal might
be, however, there could be little doubt that many others like the
Laphams would continue to aspire to Back Bay residence and the social
blessings it promised to present to them.

It was perhaps natural that this role of the Back Bay as a principal
attractive force behind the social and residential mobility that was
changing the face of modern Boston should at the same time evoke
particular negative pronouncements. To those who still defended the
traditional social values of ancestry and upbringing that had made the
West End and Beacon Hill the pride of Boston society, the Back Bay
stood as a physical manifestation of the unfortunate rise to dominance
of strictly financial indicators of an individual's worth. This view of the
Back Bay, for instance, will be found in James' *The Bostonians*;[8]
Howells gave vivid landscape expression when he described the varying
reactions to the stench that arose from the filled lands that now formed
the Back Bay district:

> People who had cast their fortunes with the New Land went by
> professing not to notice it; people who still 'hung on to the Hill' put
> their handkerchiefs to their noses and told each other the old terrible
> stories of the material used in filling up the Back Bay.[9]

In general, however, effective criticism of the Back Bay in the literature
of late-nineteenth-century Boston was not widespread. At a time when
a new aristocracy of wealth was being widely recognised and praised,
the views of Back Bay critics were essentially anachronistic. Most
authors were grandly appreciative of the Back Bay as the pre-eminent
example of social and economic segregation in the residential structure
of Boston. They were generous in their respect and praise for residents
of that exclusive district and they were generally strong in their
disdain for the less fortunate outsiders who sought to live there but
who almost inevitably failed to attain their goal.

The Passing of Tradition in the West End

The West End, dominated by the mass of Beacon Hill, suffered from a seriously split image. Historically, the West End had been the location of the most prestigious residences in Boston, a place of considerable importance in the culture of the city. This traditional meaning of the West End was frequently admired by Boston novelists, even as changes were taking place that thoroughly diminished the former glories of the district.[10] The persistence of remnant descendants of Boston's finest old families in their comfortable, dignified homes on Beacon Hill symbolised an earlier era of calmer social order no longer seen elsewhere in late-nineteenth-century Boston and very likely gone forever. As authors struggled to make sense out of the new urban scene they seemed to find solace in the relative stability and unquestioned grace of the old West End landscape.

The qualities of this image appeared in page after page of *Miss Theodora: A West End Story*, a short novel by Helen Reed which dealt as much with the West End as a place as it did with its principal character, the maiden aunt of a fine old Boston family. To Miss Theodora, a staunch defender of early American and Boston traditions, the West End was 'the real Boston'; she could never fully accept the changing social geography of the city as older families left the Hill for the more modern and fashionable Back Bay.[11] Though for a time Miss Theodora resisted these apparently inevitable adjustments, in the end she felt obliged to abandon not only the changed West End but all of Boston as well:

> She left Boston with the less reluctance, perhaps, because of certain changes — some persons called them 'improvements' — that had begun to appear in her well-loved West End. The tall apartment houses which had begun to creep in even before she left the city, the electric cars now dashing through Charles Street, were innovations that cut her to the heart.[12]

Miss Theodora left behind a strong feeling that along with these landscape changes something precious and probably irreplaceable had been lost from the quality of life in Boston.

Any attempt to come to terms with the social patterns in the new Boston required that authors recognise this breakdown of traditional

social order in the once-fashionable West End. Changes here were
particularly difficult to ignore because they openly defied the West
End designation which had until recently been the undisputed pride of
proper Boston society. Within the novels two general classes of new
residents were recognised: one a gathering of artists and writers, and the
other a clustering of Negroes and other less well-defined ethnic
groups.[13] Upon recognising these changes, however, authors were then
faced with the difficult task of comprehending them. Some chose only
to emphasise the sad passing of cherished traditions; others saw more
clearly in the new social geography of the West End the arrival of a new
social order, one with its own implications of strength and weakness. In
this time of recent change, the same landscape could at once represent
what was missing from the past and what was actually there at the
present moment.

The Dubious South End and the Boarding-house Districts

The South End and its population were cast in an extremely doubtful
light by late-nineteenth-century Boston novelists. This district was home
to many commonplace characters who were indicative of the less
successful segment of the middle class, along with an elusive collection
of individuals known by such pejorative terms as 'quacks', 'fortune
tellers', or 'transients'.[14] This population occupied a place somewhat
below the bounds of respectable society in the view from the nearby
Back Bay, a situation that contributed to a rather dubious reputation in
Boston fiction for the South End as a place to live.[15]

The inadequacies of the South End and its residents were clearly
brought out in images of the 'awful gulf' that was 'the debatable
ground between the select regions of Back Bay and the scorned
precincts of the South End'.[16] In *The Sentimentalists* by Arthur
Stanwood Pier this borderline zone became home to a family who had
aspired to the Back Bay, but whose fortunes had tumbled with a bad
investment. The depth of their failure was brilliantly symbolised by
Pier's description of their new neighbourhood:

> In this region the streets are flat, treeless, asphalted wastes, lined
> with brick shells, in most of which the vestibules bear a perforation
> of electric buttons and suggest the but recently abated presence of a
> slovenly scrubwoman. The window-curtains are uniformly of flowsy
> lace . . . On one side of a street may be a series of houses with plaster
> fronts, from which the plaster is scaling, giving them an air of suffer-
> ing from skin disease; opposite them may be houses of such plausible

exterior that one, knowing the neighbourhood, is bound to distrust the plumbing. The district is peopled largely with those who board . . . by hard-working artisans and their families . . . by persons who range from the acme of the commonplace to the abominable of Bohemia, and by clerks and professional men, whose ambition has faded, year by year . . .[17]

The presence of large numbers of boarding houses with their strong implications of transience and lack of success added to the doubtful reputations of districts such as South End. The more established segment of the urban population viewed rootlessness as a potential threat to public morality, as illustrated by the crusading Reverend Morgan's description of one locale where

Three fourths of these lodgers are females, half of them having no visible occupation . . . This street is notorious in the annals of crime. Yet Boston has many such streets. Here are miles on miles of lodging-houses instead of homes. Thousands of young men have only a private room, without fire, destitute of all endearing charms . . . with no fond ties of affection.[18]

Questions of conventional morality were serious enough in their own right to cast doubt over the desirability of these regions, but they were compounded by the fact that lack of success was itself an offence against middle-class values. There was no room for a complacent, unsuccessful, lower-middle-class population in a city where an active commitment to the struggle for mobility and wealth was one of the few cohesive social forces in an otherwise highly fragmented situation.

The Awful Tenement Districts

Relatively few of the Boston authors ventured out into the tenement districts, those regions near the heart of the city, like the congested old North End, that were home to recent immigrants and the poor. On the rare occasions when tenement districts did appear in the social geographies of late-nineteenth-century Boston, the prose conveyed an appropriate tone of shock over the scenes of human misery and environmental evils found there, suggesting that neither the authors nor their audience were at home with the realities of these landscapes. This was evident in Edward Bellamy's *Looking Backward* when the narrator awoke from his utopian dreamland to find himself in the midst of a squalid and degraded tenement district:

> From the black doorways and windows of the rookeries on every
> side came gusts of fetid air. The streets and alleys reeked with the
> effluvia of a slave ship's between-decks. As I passed I had glimpses
> within of pale babies gasping out their lives amid sultry stenches, of
> hopeless-faced women deformed by hardship, retaining of woman-
> hood no trait save weakness, while from the windows leered girls
> with brows of brass. Like the starving bands of mongrel curs that
> infest the streets of Moslem towns swarms of half-clad brutalized
> children filled the air with shrieks and curses as they fought and
> tumbled among the garbage that littered the courtyards.[19]

Perhaps the strongest condemnation of the tenements appeared in
Hamlin Garland's *Jason Edwards*, where passage after passage dwelt
upon the horrors of the physical landscape and the associated social
dangers. The neighbourhood of 'Pleasant Avenue', a name 'ironically
bitter . . . like many another in Boston', was one of

> four-storey brick and wooden buildings, rising like solid walls on
> each side of the stream of human life which filled the crevasse with
> its slow motion. Children, ragged, dirty, half-naked and ferocious,
> swarmed up and down the furnace-like street, swore and screamed in
> high-pitched, unnatural, animal-like voices, from which all childish
> music was lost.
>
> Frowzy women walking with a gait of utter weariness, aged
> women, bent and withered, and young women soon to bring other
> mouths and tongues and hands into this frightful struggle, straggled
> along the side-walks, laden with parcels, pitifully small, filled with
> food.
>
> Other women and old people leaned from open windows to get a
> breath of cooler air, frowns of pain on their faces, while in narrow
> rooms foul and crowded, invalids tortured by the deafening screams
> of the children, and the thunder of passing trams and cars, and unable
> to reach the window to escape the suffocating heat and smell of the
> cooking, turned to the wall, dumbly praying for death to end their
> suffering.[20]

As the story unfolded, the reader was constantly reminded of the horrors
of the Edwards' tenement block: the odours and noise, the few good
people and the many who were dreadful, the heat and the overcrowding

and the evil influences acting on the children. One of the characters, encountering 'a hot, unwholesome alley . . . horribly ugly . . . badly lighted . . . poison-tainted, vice-infested', reflected that the city was 'predominantly of this general character', and that this 'real' Boston did not 'get itself photographed and sent around the country'.[21]

Garland's attempts to empathise with the tenement residents earned him a thinly veiled barb in *The Complaining Millions of Men* by Edward Fuller, another author who gave more than just a little attention to the slums. Fuller questioned Garland's realism when he introduced into his own novel a fictional author, 'Hamilton Wreath' (Hamlin Garland?), overheard trying to defend the methods he employed in writing a 'sociological' novel entitled *Jared Evens* (*Jason Edwards*?):

> 'Oh, I walked up and down a place called Arragon Street a half an hour, to get the local colour . . . I tried to fancy what lives those people must lead – what wretched lives, full of misery and filth and sin. Ah! ah!', wailed Mr. Wreath, lifting his eyes to the ceiling, 'my heart bleeds for them · · it does indeed.'[22]

Garland's novel itself did little to dispel such criticism, for the tenement scenes of *Jason Edwards* were rather artlessly presented, immersed in hyperbole and redundancy. Yet it is worth noting that even Fuller seemed to agree with Garland's picture of the tenements as places of extreme poverty, degradation, and hopelessness lying far beyond the comprehension of middle-class Boston.[23]

Perhaps it was the ill fate awaiting children of the tenements that seemed to weigh most heavily on these authors. Howells may only have been playing with words when he noted the aristocratic Bromfield Corey poking fun at charity work, and referring to 'the thousands of poor creatures stiffling in their holes, and the little children dying for wholesome shelter'.[24] But there was no mistaking the meaning of a child's death in Hezekiah Butterworth's strongly anti-urban novel *Up from the Cape*. The reader's sympathy was engaged for an unfortunate newsboy who lived in North Street, 'a locality filled with tippling shops and dens of vice'.[25] Following a brief illness, the boy succumbed to the unhealthy environment, gaining at last his freedom from the slums via a poignant funeral procession out beyond the city limits to a suburban cemetery. As for the mourners, they had to return 'back to the city . . . to the wilderness of homes', a fate worse than death in a Boston dominated by greed and false values.[26]

Suburban Landscapes: Gardens of Delight or Dullness

Although the suburbs seldom played any major part in the Boston
novels, their presence just over the horizon was apparent throughout
the literature. The theme of the peaceful home in the garden received
its due share of attention.[27] Thus one Back Bay resident asked her
husband:

> 'Can't we *ever* have a house of our own . . . out in Brookline, or
> Milton, or somewhere? Even if its only a little one-storey house, I
> don't care, just so I can have a garden to plant things in, and watch
> them year after year.'[28]

The attraction of suburban parks and drives also came up time and
again in the novels, providing city residents with an opportunity for
thought and reflection away from the everyday reality of the dense
urban scene. This release was open to people of various social classes,
whether it was a carriage ride to Brookline for the well-to-do Laphams
as they discussed the need for their daughters to marry into proper
society, or a horse-car ride to Franklin Park in Roxbury for a young
couple from the tenements as they discussed the social injustices of
modern life.[29] So important were these visits to the open spaces
surrounding Boston, that they seemed to possess a symbolic power to
release both the living and the dead from the oppressiveness of urban
life. Following the death of the poor newsboy in *Up from the Cape*, his
body was borne away from the evil tenements of the city to a burial
place in suburban Mount Hope cemetery, passing first through the city,
'then away through streets where untold rural beauties mingled with
the embellishment of art . . . [which] made the blooming earth appear
like the very borders of a better land'.[30] In death, he had reached the
promised land, and it was the suburbs!

Not everyone agreed that the suburbs were superior to the city,
however. Strong adherence to what can be termed urbane values led
some authors to be rather critical of suburban locations and landscapes.
This was evident in the response of the Back Bay husband to his wife's
pleadings, quoted above, for a new home in the outlying districts:

> He made a slight, though honest, effort to reason himself into
> acquiescence with the request, but suburban life had always been
> peculiarly distasteful to him. Visions of time-tables, rainy nights,
> snow-blocked streets, sick carriage horses, and obstreperous
> grooms rose in a grimacing swarm. No! he couldn't do it. It was too
> much of a sacrifice.[31]

This critical view of the relative inaccessibility of the suburbs represented something more than physical distance to some Bostonians. There seemed to be an intangible barrier that stood between Boston and even its closest suburbs, making them appear to be farther from the city than the actual linear distance. This was illustrated by a conversation in Howells' *April Hopes* when one person observed that it was a 'great way' to Cambridge from Boston and another replied, 'Yes it *is* a great way. And a strange thing about it is that when you're living here it's a good deal further from Boston to Cambridge than it is from Cambridge to Boston.'[32]

The suburban landscape itself also came in for some harsh criticism in Boston fiction. One particularly dreary description appeared in Pier's *The Sentimentalists* as two of his characters searched throughout a new suburban community while giving some thought to purchasing a home there:

> At the foot of the hill . . . was a flat map of frame houses, simple in architecture as those a child draws on a slate, and divided into squares by narrow streets.
>
> 'Don't they look like nasty peaked little fungi, Frances? They make me want to reach out and pick one and mash it underfoot.'
>
> 'There are trees,' said Virginia, rousing herself to an effort at cheerfulness.
>
> 'Yes,' replied Frances, with a glance at a bare and mediocre elm, 'there are trees.'
>
> They passed three or four houses which were vacant and on whose plots of ground were set the T-shaped signs of the real-estate dealer.
>
> 'But Virginia, this is so pathetic! Such dreary people as they must be out here.'[33]

Indeed, the couple could not forgive the suburbs for lacking the urbanity and prestige which they associated with Back Bay, the choice residential location that remained far beyond their financial means. Some of the Boston novelists seemed to sense that even the mushrooming suburbs might lose their appeal, along with virtually all of Boston outside Back Bay, if people insisted on evaluating the desirability of their residential districts only on the basis of the social status each had to offer.

Conclusion

The social-geographic images of the Boston novelists described above presaged an uncertain future for the quality of life in the metropolis. Authors recognised many fundamental changes in the residential landscape, yet they took surprisingly little comfort in it, for they saw everywhere the potential for dissatisfaction with the urban landscape among the society that inhabited it. Apparently the middle classes were coming to dominate a way of life which their values of mobility and status might never allow them to fully appreciate. This was brought to light perhaps most convincingly by Henry James. Whereas the outskirts of Boston – the Wentworth home in *The Europeans*, for example, or his short-story settings – represent the pastoral ideal of an earlier generation, his disenchantment with the nature of modern mass urban life is fully revealed in *The Bostonians*. In this work the characters were 'sick of the Back Bay', with all its pretentiousness, yet they also 'mortally disliked' the commonness of the South End and found little joy in the sterility of the 'desolate suburban horizons'.[34] The fault lay in the very concept of the mass city, with its lack of concern for the individual and for established social order. No matter that the more intimately scaled city which James apparently preferred was probably lost forever, his artistic freedom still allowed him to lay bare the faults of the present while offering little or no hope for the future.

In this way, James and the other Boston novelists reviewed here provide us with some of the earliest examples of the potential for disaffection that came to characterise the middle-class urbanite of the twentieth century. What they suggest is that the lack of strong middle-class identification with the modern city was not an unfortunate side-effect capable of adjustment given an intelligent remedy, but rather an almost necessary outcome of such key middle-class values as those regarding status and mobility. It is a sobering thought to ponder whether the establishment of a satisfactory relationship between the middle class and the modern city may have been precluded by the inherent nature of the middle-class society itself. This is undoubtedly the most profound implication of the social-literary geography of late-nineteenth-century Boston.

Notes

1. Henry James, one of the most important authors to be considered here, was actually not a resident of Boston during this time but, rather, an expatriate living in London. However, he was a native Bostonian who retained close ties with the literary world, and his credentials for commenting on the Boston scene are seldom questioned.

2. Conceptual and methodological considerations surrounding the use of urban fiction in geographic research are discussed in W. J. Lloyd, 'Landscape Imagery in the Urban Novel: A Source of Geographic Evidence' in G. T. Moore

and R. G. Golledge (eds.), *Environmental Knowing: Theories, Research and Methods* (Dowden, Hutchinson & Ross, Stroudsburg, Pa., 1976), pp. 279–85.

3. Sylvester Baxter, 'Howells' Boston', *New England Magazine*, New Series, vol. 9 (1893), p. 130. See also Lindsay Swift, 'Boston as Portrayed in Fiction', *Book Buyer*, vol. 23 (1901), pp. 197–204; and Frances Carruth, *Fictional Rambles in and About Boston* (McClure Phillips, New York, 1902).

4. The urban geographic ideas found in the full range of urban literature are discussed in depth in W. J. Lloyd, 'Images of Late Nineteenth Century Urban Landscapes', unpublished PhD dissertation, University of California, Los Angeles, 1977.

5. Arlo Bates, *The Philistines* (Ticknor, Boston, 1899), p. 96; Margaret Allston (pseudonym of Anna F. Bergengren), *Her Boston Experiences* (L. C. Page, Boston, 1900), p. 65; Arthur Stanwood Pier, *The Sentimentalists* (Harper, New York, 1901), p. 166.

6. William Dean Howells, *The Rise of Silas Lapham*, 1884 (Riverside Press, Cambridge, Mass., 1937); William Dean Howells, *A Modern Instance*, 1881 (Houghton Mifflin, Boston, 1909); Edward Fuller, *The Complaining Millions of Men* (Harper, New York, 1893); Pier, *Sentimentalists*.

7. Howells, *Silas Lapham*, p. 33.

8. Henry James, *The Bostonians*, 1886 (2 vols., Macmillan, London, 1921), vol. 1, pp. 40 and 197.

9. Howells, *Silas Lapham*, p. 43.

10. Arlo Bates, *The Puritans* (Houghton Mifflin, Boston, 1898), p. 3; Howells, *Modern Instance*, p. 231; Howells, *Silas Lapham*, p. 194; Helen L. Reed, *Miss Theodora: A West End Story* (Richard G. Badger, Boston, 1898).

11. Reed, *Miss Theodora*, p. 1.

12. Ibid., p. 248.

13. The West End residences of the artists and writers appear in Arlo Bates, *The Pagans*, 1884 (Houghton Mifflin, Boston, 1889), p. 267; William Dean Howells, *April Hopes* (Harper, New York, 1888), p. 281; and Howells, *Modern Instance*, pp. 244–5. The Negro district appears in Reed, *Miss Theodora*, pp. 74–81; and Allston, *Her Boston Experiences*, p. 111.

14. William Dean Howells, *The Undiscovered Country*, 1880 (Houghton Mifflin, Boston, 1892), pp. 1–2; Henry Morgan, *Boston Inside Out! Sins of a Great City! A Story of Real Life*, 12th edn (Shawmutt, Boston, 1880), pp. 238–9; Maria L. Poole, *Roweny in Boston* (Harper, New York, 1892), p. 59; Fuller, *Complaining Millions*, pp. 158–9; James, *Bostonians*, pp. 31–8.

15. F. Marion Crawford, *An American Politician* (Houghton Mifflin, Boston, 1885), p. 219; Allston, *Her Boston Experiences*, pp. 68–9; Howells, *Silas Lapham*, p. 23.

16. Bates, *Philistines*, pp. 96 and 212.

17. Pier, *Sentimentalists*, pp. 409–10.

18. Morgan, *Boston Inside Out!*, pp. 100–1. See also Walter L. Sawyer, *A Local Habitation* (Small, Maynard, Boston, 1889), pp. 5–6; Pier, *Sentimentalists*, p. 410; and Fuller, *Complaining Millions*, pp. 158–9 and 226.

19. Edward Bellamy, *Looking Backward, 2000–1887* (Houghton Mifflin, Boston, 1889), p. 323.

20. Hamlin Garland, *Jason Edwards: An Average Man*, 1891 (Appleton, New York, 1897), pp. 24–7; see also pp. 30–58.

21. Ibid., pp. 90–1.

22. Fuller, *Complaining Millions*, pp. 188–90.

23. Ibid., pp. 15 and 53.

24. Howells, *Silas Lapham*, p. 201.

25. Hezekiah Butterworth, *Up from the Cape* (Estes and Lauriat, Boston, 1883), pp. 207–11.

26. Ibid., p. 211.

27. Sidney McCall (pseudonym of Mary M. Fenollosa), *Truth Dexter* (Little, Brown, Boston, 1901), pp. 156 and 233; Garland, *Jason Edwards*, p. 33; Fuller, *Complaining Millions*, p. 53; Allston, *Her Boston Experiences*, p. 66.

28. McCall, *Truth Dexter*, p. 258.

29. Howells, *Silas Lapham*, pp. 242–6; and Fuller, *Complaining Millions*, p. 102. See also Butterworth, *Up from the Cape*, p. 168.

30. Butterworth, *Up from the Cape*, pp. 208–9. The theme of rebirth in the suburbs also appears in McCall, *Truth Dexter*, pp. 277–9.

31. McCall, *Truth Dexter*, p. 236. See also Howells, *Modern Instance*, p. 245; Allston, *Her Boston Experiences*, pp. 55–6; James, *Bostonians*, vol. 1, pp. 17, 137, 210 and 263; vol. 2, p. 19.

32. Howells, *April Hopes*, p. 359.

33. Pier, *Sentimentalists*, pp. 251–3.

34. James, *Bostonians*, vol. 1, pp. 34, 79 and 210.

11 NINETEENTH-CENTURY ST PETERSBURG: WORKPOINTS FOR AN EXPLORATION OF IMAGE AND PLACE

Howard F. Andrews

In 1910 Prince Peter Kropotkin concluded an article on 'St. Petersburg' for the eleventh edition of the *Encyclopaedia Britannica* as follows:

> It has often been said that St. Petersburg is the head of Russia and Moscow its heart. The first part at least of this saying is true. In the development of thought and in naturalizing in Russia the results of west European culture and philosophy St. Petersburg has played a prominent part. It has helped greatly to familiarize the public with the teachings of west European science and thinking and to give to Russian literature its liberality of mind and freedom from the trammels of tradition. *St. Petersburg has no traditions, no history beyond that of palace conspiracies and there is nothing in its past to attract the writer or the thinker.*[1] (My italic.)

Kropotkin's personal experiences in St Petersburg do not seem to have been especially fulfilling and his largely negative views on the administration and social organisation of the large nineteenth-century city may have contributed considerably to his somewhat jaundiced perception of the Russian capital.[2] Nevertheless, it is difficult *not* to take issue with his final conclusion: Kropotkin's condemnation of the city notwithstanding, numerous writers and thinkers of considerable stature were attracted to St Petersburg both physically and figuratively, and it plays a prominent part as setting and milieu, antagonist and protagonist, in a number of writings in nineteenth-century Russian literature. While many of these writings were discussed with insight and sensitivity by Kropotkin in his series of lectures at Boston's Lowell Institute in 1901, the 'St. Petersburg connection' receives only passing mention, and may be construed as a vehicle for intimating his wider concerns for the general structure of society and the individual's daily conduct of life.[3]

This chapter is a brief attempt to explore the treatment of St Petersburg by selected nineteenth-century writers. Its subtitle emphasises the tentative and preliminary nature of such an undertaking, an initial pass, as it were, through one kind of source material and information which may be used to inform a full-scale expedition. Major points of relief may be indicated and a number of the more striking routes for the traveller may be suggested, but the delicacy of the landscape and the intricacies of its topography are only barely intimated and remain largely hidden from view. The grain of the country is unmistakable, however. Thus, the common theme running through all of the writings discussed here is the existence of St Petersburg as a *place*, a tangible reality of certain memorable and describable elements, and as an *image*, a state of mind adduced by place which is admittedly less tangible but no less real.[4] Even in this apparently simple characterisation of the myriad themes both explicit and implicit in the literature, there resides a profound paradox: had T. S. Eliot not used the phrase 'Unreal City' for a different place and time, it could have been applied with equal ease and force to St Petersburg.

The City in Context

Some background information about the city may be summarised briefly at this point, though to understand fully the reasons for its foundation would require a full-length study of the life and times, and particularly the psyche of Peter the Great.[5]

The coastal strip of the Gulf of Finland was an early battlefield between Russia and Sweden in the Great Northern War (1700–21) during which Russia finally obtained command of the Baltic. During the campaigns of 1702–3, the 'key city' of Noteburg was captured and renamed Schlusselburg by Peter the Great. A little distance away towards the mouth of the River Neva, the new Russian fortress of St Peter and St Paul was built (1703) and beyond the walls of the fortress, 'on a desolate stretch of wind-blown marsh, Peter laid the foundations of his city, his future "paradise", St. Petersburg, Russia's window on the west'.[6] There followed a period of incredible activity in building the city, during which tens of thousands of lives were lost through recurring epidemics amongst the workers, herded to the swamps of the St Petersburg site from every part of Russia. Westerly and south-westerly winds off the Gulf of Finland played havoc with the normal flow of the Neva, driving back the waters and causing innumerable and disastrous floods: indeed in the 250 years since 1700 the Neva overflowed its banks more than 300 times.[7] At the earliest opportunity Peter transferred the court and seat of government from

Moscow to his new city, which was to stand for 'the tutelary light of
the west against the Byzantine dark of Moscow'.[8] A year before Peter's
death in 1725, the relics of the Russian warrior-saint Alexander Nevsky
were transferred from the ancient capital of Vladimir via the former
capital of Moscow to the new capital: for practically two centuries the
new city was to symbolise the growing force of Russia in European
affairs besides the manifestation of Tsarist autocracy. Only rarely can
we find such a dramatic example of the imprint on the landscape of a
single individual's decisions.

Subsequent rulers brought changes to the design of the city and the
lives of its inhabitants. Under Anna, during the 1730s, a rigid German
Baltic bureaucracy developed in St Petersburg along the lines laid down
previously by Peter the Great. Under Elizabeth (1741--61) great new
buildings were constructed in the city, among them the famous Winter
Palace, with which she helped to create and maintain something of the
illusion of another Versailles in St Petersburg.[9] This 'illusion' almost
became reality under Catharine the Great (1762–96), the golden age of
Russian opulence and glitter. For the privileged enclave, St Petersburg
was the centre of an extravagant and ostentatious lifestyle supported by
wealth drained from enormous estates.[10] This period also, and not
unexpectedly, witnessed Russian territorial aggrandisement on a scale
unparalleled since Ivan the Terrible.

It was Catharine the Great who commissioned the French sculptor,
Falconnet, to work on the famous equestrian statue of Peter the Great,
finally unveiled in 1782 after some twelve years' labour. Writing to
Diderot at the beginning of his work, Falconnet remarks:

> I shall limit myself to a statue of a hero and I shall not portray him
> as a great commander and conqueror, though of course he was both;
> far greater is his personality as a creator, a legislator, a benefactor to
> his country and that is what ought to be shown.[11]

The statue portrays a magnificent charger rearing at the edge of a huge
rough-hewn rock: its front hooves are poised in mid-air as if to begin its
ascent and its hind hooves trample the head of a vanquished and
twisting snake. Astride the horse is Peter the Great, bridle in one hand
and the other outstretched. This gesture 'seen from the right looks
threatening, but viewed from the left, can be interpreted as benevolent,
almost benedictory'.[12]

The various themes which emerge from this short background
statement occur repeatedly in the creative writings of a number of

nineteenth-century authors. Of these, Pushkin, Gogol, Dostoevsky and
Bely are by far the best known and provide us with major sources for
an exploration of image and place, of the geography of imagination.
Workpoints concerning each author follow.

The City in Nineteenth-century Literature

Pushkin

It is difficult to find an English language survey of Russian literature in
which Pushkin is *not* heralded as the father of Russian literature, and
likewise, the major ranking authors of the nineteenth century have all
at some time acknowledged their debts to him. Apart from less well-
known figures,[13] Pushkin is the first to write on the theme of St
Petersburg and Peter the Great, and Falconnet's famous statue is
probably better known as *The Bronze Horseman*, the title of Pushkin's
last great narrative poem (written in 1833, published posthumously in
1841). D. S. Mirsky has written the best summary of the poem for our
present purposes:[14]

> Its philosophical . . . subject is the irreconcilable conflict of the
> rights of the community, as incarnate in the *genius loci* of the city,
> the bronze statue of Peter the Great on the Senate Square — and
> those of the individual as represented by the wretched Evgeny, who
> is undone by the mere geographical factor of the site of Petersburg.

'The mere geographical factor'! The central span of attention in the
poem concerns the impact of the disastrous Neva flooding in St
Petersburg (1824) on the life of Evgeny, an insignificant young civil
servant. On the night of the flood, Evgeny is marooned at the corner of
Senate Square where the equestrian statue of Peter stands riding the
flood waters and he is panic stricken over the safety of his loved one
who lives on one of the islands in the Neva. As the flood abates he is
rowed across to the island to find both she and her home have been
washed away. Evgeny finally goes mad over his loss. A year later finds
him destitute and roaming the streets of the city. On a wild stormy
night he is reminded again of the flood and his lost love, and finding
himself before the statue of Peter the Great, associates the cause of his
loss and misery with the autocrat who founded the city. As he curses
the statue — echoing the legendary curse of Peter's first wife, Avdotia
('Let Petersburg be empty!')[15] — his feverish mind brings it to life and it

threatens him with outstretched arm. Terrified, he flees and hears the bronze horseman in pursuit, 'galloping with heavy clatter' down the misty Petersburg streets. At the end of the poem his body is found close to the remains of his loved one's home on the island.

Throughout the entire poem there is carefully sustained a worship of the city and its founder, contrasted dramatically with the misery of the wretched Evgeny: 'the pitiful citizen . . . had no more right to question the will of the bronze figure than had the toiling masses who gave their lives that Peter's great city might exist', writes E. J. Simmons in summarising this conflicting theme in the poem.[16] The issue of the moral conflict remains unsolved, but the sense of place instilled in Pushkin's lines – Mirsky's *'genius loci'* – is a theme that reverberates through literature for almost a century after.[17] Indeed, it has been claimed that Pushkin 'is in the same measure the creator of the *image* of Petersburg as Peter the Great was the builder of the city itself'.[18]

Gogol

It could further be said that if Pushkin created the image of Petersburg, then the major contribution of Gogol was to the elaboration of its atmosphere. The cycle of stories known as the *Petersburg Tales* together project onto the real city a sense of the bizarre, the fantastic and perhaps also the grotesque.[19] Again, some knowledge of the extraliterary circumstances surrounding St Petersburg – called by Dostoevsky 'the most fantastic city with the most fantastic history of any city on the globe'[20] – is required to begin to approach an understanding of the extraordinarily strong spirit of place it possesses:

> At the heart of the myth of Petersburg is the image of an unreal city, an image countenanced historically by the fact of the city's founding as an arbitrary act of will (thus, again, Dostoevsky's designation of it as 'the most abstract and intentional city in the world') and countenanced physically, so to speak, by the peculiar Petersburg situation and climate.[21]

In Gogol also we find a writer concerned with accounts of the day-to-day happenings, the mundane, pettifogging existence of the lower levels of the bureaucratic hierarchy and an acute eye for the unwritten social rules restraining interaction between classes, occupations and different levels of the government service.[22] In this regard, writes Donald Fanger, Gogol's main talent is 'in the vivid exhibition of the ordinary in characters and in their lifestyles, an ordinariness which . . .

constitutes a special challenge to the artist who would endow its representation with high poetic value . . . never had character and milieu been so intimately and totally expressive of each other'.[23]

As the setting of such diverse and innumerable distinct social groups, Petersburg lacks unity and coherence. Defined in terms of its elusiveness and lack of positive unity, the city assumes a strongly *negative* image: 'the city that lacks a heart becomes a creature of heartlessness'. Also, apart from the writer's correspondence, broad descriptions of the city itself are rarely given by Gogol and where they do occur, they are concerned with the pitiful quality of life of the experiencing individual rather than the panoramic view or vista of the urban structure. In such a context exists the well-known character of Akaky Akakievich in the tale of 'The Overcoat': 'a pathetically humble and inferior figure'[24] who seems to be almost totally at the mercy of the circumstances of the city itself. This perspective of the individual totally subordinate to his background is what Rahv has used to describe naturalistic writing[25] — a situation in which the city itself assumes the role of hero (or anti-hero in the case of Gogol's St Petersburg) in place of its inhabitants. In fact, Fanger points out, Gogol not only makes his characters subordinate to the city, he appears more frequently to debase them in evoking a myth of St Petersburg whose peculiar atmosphere diffuses terror and alienation.[26]

This last point is critical in coming to grips with the spirit of place pervading Gogol's *Petersburg Tales* and in locating his contribution to what Fisher would dub 'literature of the city' as distinct from 'literature about the city'.[27] Together with Pushkin's *Evgenii Onegin* and Lermontov's *A Hero of Our Times*, Gogol's *Dead Souls* is frequently identified as heralding the emergence of the golden age of Russian literature.[28] He was also most clearly connected to the development of literary realism and the so-called natural school in which the 'social novel' genre became dominant, and it is this particular connection which formed the primary critical frame of reference for evaluating and interpreting his work for most of the nineteenth century. Distinct from 'Gogol as civic realist', or 'Gogol the Russian Dickens', however, there emerged an understanding of his work in the stream of symbolist criticism and thought. It is from this latter perspective that Gogol is arguably the first major contributor to urban literature in which the city is viewed not as 'something out there' to be objectified, dissected and described in detached and minute detail, but instead as a mode of thought and being, a context which persuades and structures consciousness and experience of the world. Side by side with the

realism and acutely observed detail of Gogol's *Petersburg Tales* sits a
parallel surrealism, most evident, for example, in the famous tale 'The
Nose'; the latter is not simply a bizarre contradiction of the notion of
reality, but is principally a different conception of reality, driven by a
distinctive 'realistic' mode of illusion that is profoundly urban in its
genesis. A senseless world, the apparently bizarre, the truly fantastic,
the grotesque, the sense of evil presence are all peculiar traits of this
modality of experience which, as Fisher has noted, are articulated in
continuous fashion with modernist writers like Joyce, Rilke, Kafka and
Eliot.[29] Andrey Bely, discussed below, is clearly within this same
category: Gogol, it may be argued, preceded them all. As Remizov
approached it, for Gogol, 'the world is a witches' sabbath; whether
asleep or awake man is forever under an evil spell; and there is nothing
to wake up into'.[30]

In Gogol's case it is possible to detect something of the evolution of
this sense of St Petersburg as place through his surviving correspondence.
Correspondence reflecting his youthful anticipations of life in the city
and his initial experiences are in sharp contrast to the reactions and
moods expressed in letters home within weeks of his arrival: early thrill
and awe, bedazzlement and wonder were quickly replaced by disillusion-
ment and depression.

> A most extraordinary silence reigns over the city: one cannot scent
> even the slightest whiff of spirit in anyone: everyone works in an
> office and talks of administrative questions and relations with
> colleagues and nothing else: everything is stifled, enmired in the
> infinitesimal preoccupations and petty labors that constitute the
> sterile lives of these people.[31]

Dostoevsky

Of the four writers being discussed here, Dostoevsky has without a
doubt received the most critical attention and the volume of material to
be sifted and analysed in the present context is simply enormous.[32]
Further, St Petersburg occupies such a central position in his writing (at
least two-thirds of his novels, stories and sketches are set in that city)
and there exists such an extensive literature dealing explicitly with this
aspect of his work, that it would be a frustrating and certainly daunting
prospect to attempt to make a decent approach to the topic within the
limits of this brief survey.[33] 'Nowhere else, in his opinion, did the *genius
loci* produce so dreadful a fever of the soul, such fantastic, and at the
same time abstract, visions and ventures of delirium.'[34] And having

been infected with this powerful *genius loci* of St Petersburg, this
'magical vision, a dream',[35] Dostoevsky will always be remembered for
the effectiveness with which he imparted this sense of place:

> there is no doubt that he knew how to create a Russian landscape
> and to people it with Russians. Reading Dostoevsky, we smell it and
> taste it; his Petersburg sticks to us just as Dickens' London does and
> we will never be free of it again.[36]

> Nineteenth-century Petersburg, in spite of all the fantastic colouring
> Dostoevsky imparted to its descriptions has not been depicted by
> any one more exactly, more sharply, more palpably or more truly.[37]

> But while St. Petersburg was largely a map to Dostoevsky, a few
> incised streets where his heroes wandered, it was a map he had
> studied carefully and knew by heart . . . Dostoevsky's imaginary
> map-like St. Petersburg included a few very real places, but these
> places are made real only because concentrations of spiritual energy
> took place in them.[38]

> The streets and squares, alleyways and canals not only serve as a
> background to the action but enter with their outlines into the
> thoughts and actions of the heroes. The city constantly dominates
> the people and hangs over their fates . . . Petersburg is inseparable
> from the personal drama of Raskolnikov: it is that fabric upon
> which its cruel dialectic draws its design.[39]

As the dominant theme of these comments suggests, Dostoevsky's
contribution to the evolving image of St Petersburg has most often been
viewed in relation to the place of his fiction in the evolution of the
social novel and natural realism — romantic or otherwise. This is
particularly evident in a recent work by Kabat,[40] in which St Petersburg,
as described and portrayed by Dostoevsky in his diaries and novels, is
discussed as epitomising his obsession with the alienation and subversion
of Russia's identity through contact with European institutions and
social movements.[41] The eclecticism of the city's architecture, for
example, is seen as testimony to the city's lack of an organic identity,
to the 'fragmentation and alienation which it imposes on its
inhabitants'.[42] Seen through the eyes of Raskolnikov and other
characters in *Crime and Punishment*, Petersburg is perceived as 'a
hostile and threatening environment' summoning up 'apocalyptic and

infernal imagery'.[43] As with subsequent appraisals of Gogol's work, the field of Dostoevsky scholarship is by no means devoid of different interpretations which similarly have important implications for exploring image and place, and more especially for grasping the radically different (urban) structure of consciousness which underpins his novels and short stories.[44]

Bely

The final writer to be noted here is Andrey Bely about whom probably the least amount is available in the English language. While the literary influences between Pushkin, Gogol and Dostoevsky may be reasonably well identified in their various works, to some extent Bely stands apart from these three as a leading figure of the Russian symbolist school, though as Mochulsky notes, Bely the writer 'comes from Gogol'.[45] In many of Bely's works, the Gogolian grotesque is at the same time made more complicated and more opaque.

With an accumulated sense of how St Petersburg has been viewed or exploited in the works of Pushkin, Gogol and Dostoevsky, together with some idea of the approach of the Russian symbolists, we might conceivably predict something of the manner in which the setting of the city is handled by Bely in his novel *Petersburg* (1913). Whereas French symbolism was the celebration of a new form of poetic expression, Russian symbolism, according to Mirsky, became a philosophy.[46] The phrase 'des forêts de symboles' in the sonnet 'Correspondances' by Baudelaire took on a literal interpretation in which the significance of all artifacts or events were symbolic of something else. The distinctions between life and art became progressively blurred, until they disappeared almost entirely. 'Poems were perceived as life; life created poems.'[47] Within the Russian symbolist movement, writers like Blok, Bely and Ivanov strove to make symbolism a metaphysical and mystical philosophy: 'Bely's world, for all its almost more than life-like detail, is an immaterial world of ideas into which this reality of ours is only projected like a whirlwind of phantasms.'[48] In his second collection of poems, *Ashes*, published at the end of 1908, the verses comprising the third section, entitled 'The City', portray the sinister ambivalence and illusiveness of the modern city in which passers-by and actors in the daily round are simply masked participants in the continuing masquerade of the urban carnival. Weaving in and out between actors and observers moves the red domino, Fate, joining the themes of *Ashes* with the presence of the same elemental force in his novel, *Petersburg*.

Mochulsky quotes a long section from Bely's own *Memoirs*, in which the genesis of the novel *Petersburg* is described for us:

> On long autumn nights, I scrutinized the images swarming around me: from beneath them the central image of Petersburg slowly ripened for me. It exploded within me so unexpectedly and strangely, that I was obliged to dwell on it . . . Instead of the figure [of the Senator Ableukhov] and the environment there was something difficult to define: not a colour, not a sound; and I sensed that the image must be ignited from some sort of dim harmony . . . there burst out before me the picture of the Neva with the bend of the Winter Canal: the dim, moonlit, bluish-silver night and the square of the black carriage with the red lamp . . .[49]

The novel is set in St Petersburg in 1905 at the time of the revolution, and concerns a conflict situation between father and son, the one representing the formalism of the St Petersburg bureaucracy, the other the rationalism of the revolutionaries. 'The plot is presented as a fumble of mistakes, love, betrayal, accidents and *quid pro quos* – a true picture of the confusion and hodge-podge of reality . . .'[50] Bely creates a terrifying, unbelievable world where delirium seems the pathologically normal, unremarkable mode of experience. The grotesque of Gogol and the fantastic city of Dostoevsky pale beside this inspired horror[51] and recede before the spectral nightmare world of monstrous absurdities that he portrays. The affinities with the contemporary development of the 'Teatro del Grottesco' and the taken-for-granted horror of Kafka's urban world are strongly apparent. For Bely, the unreality of St Petersburg rests not simply in its original foundation 'as an arbitrary act of will'. The myth of Dostoevsky's Petersburg is here replaced by delirium and dream (passages from *The Bronze Horseman* precede each chapter) and the characters or actors play out their almost predetermined parts as abstract figures and mechanical automatons in an eerie, mystical atmosphere of twilight in a cold, foggy winter. They appear as 'streams of shadows', passing on and off stage as (literally) buildings and people loom in out of the mist: 'The city possesses them . . . held as it were by a hypnosis we attribute to dreams.'[52] The city and all its inhabitants are portrayed as phantoms, disembodied and dematerialised. Solid objects are disincarnated as silhouettes decomposed into 'a smear of soot' and 'shining lunar ash': 'An enormous red sun hovered above the Neva: the buildings of

Petersburg seemed to have dwindled away, transformed into ethereal, mist-permeated amethyst lace . . .'[53]

Into this twilight world in a blinding flash, the Bronze Horseman, 'ponderous greenish arm outstretched', visits the student revolutionary Dudkin:

> The Bronze Horseman greeted him, 'How do you do, son?' . . . he lowered his metal body, cast in bronze, into a chair, and let his green gleaming elbow drop with its full weight of bronze clanging onto the table. Slowly, the Emperor removed the bronze garland from his brow; the laurel leaves clanged as they fell.[54]

Mochulsky makes the clear connection between Bely's Dudkin and Pushkin's Evgeny at this point,[55] but there is an important distinction to be made. The irreverence given by Bely to the Emperor's familiar greeting ('How do you do, son?') negates the status of the Emperor — or the Tsar's censors — of Pushkin's day, and Pushkin's famous lines, 'Flaunt your beauty, Peter's city, and stand/unshakable like Russia',[56] could not project a sense of place much further from the image given through Bely's reaction.

The unreality of St Petersburg for Bely reflected the shipwreck of a culture, 'the product of empty cerebral play': its image became a perfect expression of a reality created solely by the mind.[57] Its culture was exhausted, the strength of its hold on the country an illusion. It would disappear when the (interminable) greenish-yellow fog was dispelled by a storm:

> In a great cataclysm St. Petersburg . . . will vanish from the face of the earth; and, moreover, it will vanish from the face of the earth (the Dostoevskian motif!) shrouded in a fog. For there is no St. Petersburg: 'It only appears to exist.'[58]

Conclusion

The span between Pushkin and Bely is less than a hundred years, between the foundation of St Petersburg and the October Revolution a little more than two hundred years. In such a short space of time both Russian literature and the city underwent enormous changes of lasting and powerful influence on subsequent events. The *genius loci* of St Petersburg was felt as something concrete and tangible to be wondered at in awe, admitted reluctantly or finally rejected. At all times, it seems, it was felt as some sort of presence inextricably involved in the lives and destiny of not only its residents, but of the rest of the Russian people besides. For Pushkin it was an image formed in exaltation; for Gogol, an atmosphere and fantasy. For Dostoevsky it was a myth of subtle moods; and for Bely, a ghostly nightmare composed symbolically. The physical reality of the city is thus transcended in the minds of those who experience it, supporting the view of St Petersburg as 'the most artificial city that men have constructed on the face of the earth'.[59]

If we view the Bronze Horseman as the same fundamental symbolic point of arrival and departure for both Pushkin and Bely, both authors may be seen to occupy different developmental levels of the same axis in the enduring dialectic between writer and place. The movement of this dialectic may be traced through the regenerative influences of intervening authors such as Gogol and Dostoevsky and in the profound economic and social changes affecting the fabric of the city and the wider society. Each author, with an intensely individual insight and skill, perceived particular meanings in these changes and pursued their implications in different ways. The image projected of St Petersburg as a result of this dialectic is multilayered and not available to frontal assault.

Despite the extraordinary circumstances surrounding the development of this particular city, however, and despite the enormous wealth of material available for an exploration of the spirit of this particular place, it is inconceivable that the sense of *any* place may be grasped within a single methodological frame of reference.[60] In his recent study of Gogol, Donald Fanger quotes a phrase from Max Hayward which may readily be applied to the phenomenon of spirit of place: it is, it seems, 'protected by an indeterminacy principle which excludes or frustrates any single approach'. Fanger's subsequent comment is equally telling:

> It is clear, however, that any effort to answer the question while respecting the intricacies of the phenomenon under investigation must avoid terms that might prejudge it — must become, in some sense, an exercise in applied phenomenology, where the interaction is constantly scrutinized, examined and justified.[61]

We are left, then, with a somewhat daunting prospect if we are ever to proceed beyond the present stage of sketching workpoints as prolegomena to sustained explorations of image and place, and begin the expedition proper and the large-scale mapping itself. As indicated in the introduction to this chapter, these workpoints represent little more than the beginnings of a traveller's checklist of items to be packed for a long voyage and -- to stretch an already strained metaphor still further - they represent no more than a glimpse of some of the contents of just one of many trunks to be packed. Thus, for example, there is the obvious matter of additional source materials: the correspondence, diaries, autobiographical fragments and workbooks of the writers themselves and their contemporaries; travellers' accounts and statements of those experiencing the place at different times and under different circumstances; contemporary paintings, sketches and maps; and the harder, more opaque, but certainly more familiar documentation of social science and history, reflecting the economic, social and political conditions and events of everyday life. For Fanger is surely correct in arguing the necessity of avoiding a premature definition of the problem itself, or at least in being able to retain a degree of indeterminacy that is less usual in the tradition of mainstream social science. As with Sartre's mammoth effort to unravel the biography of a single individual (Flaubert), we may with reason expect the detailed mapping of the sense of one particular place to present us with similar problems of closure. However, if we view the work of creative writers and thinkers as the cutting edge of society's continuing attempt to make sense of its own history, unfolding the image and sense of place revealed in their works against the backdrop of a constant process of social change represents as good a starting point as any and one that is potentially more fertile than most. It is, for example, in such a context that Kropotkin's comments, which opened this chapter, take on a meaning enriched beyond the particular details of his own biography and enable the voyager from the present future to regain vicariously the world of the contemporary past. It is, after all, a world we may experience only through informed imagination, discovered and rediscovered continuously as we grasp the fictive manipulations of writers and thinkers in their own efforts to comprehend and make sense of their own worlds.

Notes

1. *Encyclopaedia Britannica*, 11th edn (29 vols., Cambridge University Press, New York, 1910–11), vol. 14, p. 40 (emphasis added). The article was signed jointly with J. T. Bealby, former editor of the *Scottish Geographical Magazine*, but an almost identical passage – under Kropotkin's name only – also concluded

the 'St. Petersburg' article in the earlier, ninth edition: *Encyclopaedia Britannica*, 9th edn (25 vols., Adam and Charles Black, Edinburgh, 1875–89), vol. 21, p. 195.

2. See, for example, G. Woodcock and I. Avakumović, *The Anarchist Prince: a Biographical Study of Peter Kropotkin* (T. V. Boardman, London, 1950), pp. 30–53 and 76–7. Kropotkin's views on the city in the eighteenth and nineteenth centuries are discussed in R. W. Jones, 'The Anarchist Communism of Peter Kropotkin and its Importance for Contemporary Human Geography', unpublished MA thesis, Department of Geography, University of Toronto, 1972, pp. 100–10.

3. P. A. Kropotkin, *Russian Literature: Ideals and Reality*, 2nd edn (Duckworth, London, 1916); note, for example, the discussion of Pushkin's *Eugenii Onegin* (pp. 47–53) and especially of the works of Dostoevsky (pp. 179–87).

4. I am reminded of the conclusion to George Ivask's brief survey, 'The Vital Ambivalence of Petersburg', *Texas Studies in Literature and Language*, vol. 17 (1975), pp. 247–55: 'Whether blessed or cursed, Petersburg *is*.'

5. The major work by J. H. Bater provides a good summary of the city's early history: J. H. Bater, *St. Petersburg: Industrialization and Change* (McGill-Queen's University Press, Montreal, 1976), pp. 17–52. See also J. Blum, *Land and Peasant in Russia from the Ninth to the Nineteenth Century* (Princeton University Press, Princeton, NJ, 1961); and B. H. Sumner, *Peter the Great and the Emergence of Russia* (Macmillan, New York, 1961).

6. R. D. Charques, *A Short History of Russia* (Dutton, New York, 1956), p. 111.

7. E. Hill in A. S. Pushkin, *The Bronze Horseman*, 1841 (Bradda Books, Letchworth, 1967, edited and introductory notes by E. Hill), p. 3.

8. Charques, *Short History*, p. 116.

9. Ibid., p. 123.

10. See, for example, Blum, *Land and Peasant*, pp. 386–91.

11. Cited in Hill (ed.), *The Bronze Horseman*, p. 4.

12. Ibid.

13. For example, A. D. Kantemir, G. R. Derzhavin and C. N. Batyushkov. Batyushkov is discussed briefly in Ivask, 'The Vital Ambivalence of Petersburg', p. 249.

14. D. S. Mirsky, *A History of Russian Literature* (Random House, New York, 1958; revised edition by F. J. Whitfield, Vintage Books).

15. Noted in Ivask, 'The Vital Ambivalence of Petersburg', p. 247.

16. E. J. Simmons, *Pushkin* (Knopf, New York, 1964, Vintage edition).

17. Steiner sees the importance of Pushkin in the realms of moral protest and social criticism rather than in those of style and literary convention, an echo, perhaps, of a distinction made by A. Slonimsky in discussing Pushkin's 'historical prose and prose dealing with everyday life'. See G. A. Steiner, *The Queen of Spades* (New American Library, New York, 1961), p. viii; and A. Field (ed.), *The Complection of Russian Literature* (Atheneum, New York, 1971), p. 42.

St Petersburg has a presence in several other pieces by Pushkin. In *Eugenii Onegin* (1823–31), for example, the superficial splendour of St Petersburg society is frequently contrasted to the tedium of the country estate. *The Queen of Spades* (1834), whose central character, Hermann, has been seen by some as the prototype for Dostoevsky's Raskolnikov, is also set in St Petersburg. Dostoevsky later called Pushkin's Hermann 'a colossal figure, an extraordinary, completely Petersburg type'. See also M. Schwartz and A. Schwartz, '*The Queen of Spades*: a psychoanalytic interpretation', *Texas Studies in Literature and Language*, vol. 17 (1975), pp. 275–88.

18. N. P. Antsiferov, cited in Donald Fanger, *Dostoevsky and Romantic Realism: a study of Dostoevsky in relation to Balzac, Dickens and Gogol*

(University of Toronto Press, Toronto, 1967, Phoenix edition), p. 104. This is a remarkable work with a central focus on the spirit of the city in the writings and development of these authors. It amounts to the major existing attempt to map the terrain whose exploration is envisaged here.

19. Ibid., Ch. 4; N. A. Nilsson, *Gogol et Pétersbourg* (Almquist and Wiksell, Stockholm, 1954); D. Fanger, *The Creation of Nikolai Gogol* (Belknap Press, Cambridge, Mass., 1979), pp. 110–20; V. Erlich, 'A Note on the Grotesque – Gogol: a Test Case' in M. Halle (ed.), *To Honor Roman Jakobson* (Mouton, The Hague, 1967), pp. 630–3.

20. This is taken from Dostoevsky's *Winter Notes on Summer Impressions*, quoted in G. C. Kabat, *Ideology and Imagination: the Image of Society in Dostoevsky* (Columbia University Press, New York, 1978), p. 21.

21. Fanger, *Dostoevsky*, p. 105. Climatic symbols, details of weather and descriptions of climatic conditions are probably the most frequently elaborated themes in stories set in St Petersburg. Always, it seems, there is sleet, rain, fog, mist, wind or snow. Add to this the varying quality of light at different times of the year at this latitude and the imagination is primed for flights of fantasy. This atmosphere is exploited in a tragi-comic manner by Gogol in several of his *Petersburg Tales*, and in a dreamlike, symbolic manner by later writers such as Blok and Bely.

22. As in, for example, Gogol's article 'St. Petersburg Notes for 1836', in addition to 'The Nose', 'The Overcoat' and others from the *Petersburg Tales*. See L. J. Kent (ed.), *The Collected Tales and Plays of Nikolai Gogol* (Random House, New York, 1966). After reading these short pieces it is difficult to accept Nabokov's statement that 'Gogol's heroes happen to be Russian squires and officials; their imagined surroundings and social conditions are perfectly unimportant', cited by P. Rahv in D. Davie (ed.), *Russian Literature and modern English fiction* (University of Toronto Press, Toronto, 1965), p. 243. Note also W. B. Lincoln, 'The Daily Life of St. Petersburg Officials in Mid-nineteenth Century', *Oxford Slavonic Papers*, New Series, vol. 8 (1975), pp. 82–100.

23. Fanger, *Creation of Nikolai Gogol*, pp. 19–20.

24. Mirsky, *History*, p. 159, who goes on to say that this particular tale gave rise to a whole genre of stories about the impoverished clerk, the most important descendant being Makar Devushkin of Dostoevsky's *Poor Folk*. This particular view of Gogol's influence is questioned by Kent (ed.), *Collected Tales*, p. xxxix.

25. P. Rahv, 'Notes on the Decline of Naturalism' in J. W. Aldridge (ed.), *Critiques and essays on modern fiction* (Random House, New York, 1952).

26. Fanger, *Dostoevsky*, pp. 115–21.

27. P. Fisher, 'City Matters: City Minds' in J. H. Buckley (ed.), *The Worlds of Victorian Fiction* (Harvard University Press, Cambridge, Mass., 1975), pp. 371–89.

28. Fanger, *Creation of Nikolai Gogol*, pp. 24–44.

29. Fisher, 'City Matters: City Minds', p. 378.

30. Cited in V. Erlich, 'Gogol and Kafka: note on "realism" and "surrealism" ' in M. Halle (ed.), *For Roman Jacobson* (Mouton, The Hague, 1956), pp. 100–8.

31. Quoted in H. Troyat, *Divided Soul: the Life of Gogol* (Minerva Press, New York, 1975), p. 45; see also C. R. Proffer (ed.), *Letters of Nikolai Gogol* (University of Michigan Press, Ann Arbor, 1967); V. Nabokov, *Nikolai Gogol* (McClelland and Stewart, Toronto, 1961), pp. 10–11.

32. See, for example, the English language bibliography up to 1958 in M. Beebe and C. Newton, 'Dostoevsky in English: a Selected Checklist of Criticism and Translations', *Modern Fiction Studies*, vol. 4, no. 3 (1958), pp. 271–91.

33. As noted above, a key guide for this topic is Fanger, *Dostoevsky*.

34. V. Ivanov, *Freedom and the Tragic Life: a Study in Dostoevsky* (Noonday Press, New York, 1952), p. 73.

35. W. Hubben, *Dostoevsky, Kierkegaard, Nietzsche, and Kafka* (Collier Books, New York, 1962), p. 54.

36. A. R. MacAndrew (ed.), *Great Russian Short Novels* (Bantam Books, New York, 1969), p. 11.

37. L. P. Grossman, cited in Fanger, *Dostoevsky*, p. 129.

38. R. Payne, *Dostoevsky, a Human Portrait* (Knopf, New York, 1961), p. 203, cited in A. D. Mead, 'Dostoevsky in Detail, a Study of the Artistic and Literary Contributions of his Descriptions of Living Quarters', unpublished PhD thesis, Vanderbilt University, 1971), p. 24.

39. L. P. Grossman, 'Dostoevsky's descriptions: the characters and the city' in G. Gibian (ed.), *Crime and Punishment* (Norton, New York, 1964), p. 674.

40. Kabat, *Ideology and Imagination*.

41. Ibid., p. 21. Kabat quotes from Dostoevsky's diary in this connection: 'In these buildings one may read, as from a book, the tide of all ideas, and petty ideas, which, gradually or suddenly, have flown to us from Europe and which have finally subdued or enslaved us.'

42. Ibid., p. 120. Note also Pearl C. Niemi, 'The Art of *Crime and Punishment*', *Modern Fiction Studies*, vol. 9, no. 4 (1963), pp. 291–313.

43. J. T. Lloyd, *Fyodor Dostoevsky* (Eyre and Spottiswoode, London, 1946), p. 19.

44. In addition to Fanger, *Dostoevsky*, see an essay by L. P. Grossman, 'Composition in Dostoevsky's Novels' in *Balzac and Dostoevsky* (Ardis, Ann Arbor, 1973), pp. 54–98; and M. Bakhtin, *Problems of Dostoevsky's Poetics* (Ardis, Ann Arbor, 1973). Although Bakhtin's thesis concerning Dostoevsky as the creator of the 'polyphonic' novel is not without its critics, his discussion of 'carnivalistic' themes and characteristics in Dostoevsky's works and the identification of the antecedent literary genres informing this quality has a direct bearing on the study of literary responses to the modern city in general. See also J. Franco, 'History and Literature: Remapping the Boundaries' in W. M. Todd III (ed.), *Literature and Society in Imperial Russia* (Stanford University Press, Stanford, 1978), pp. 11–28.

45. K. Mochulsky, *Andrey Bely: his Life and Works* (Ardis, Ann Arbor, 1977), pp. 17, 75 and 120–1. Bely's last work, *The Art of Gogol*, is described by Mochulsky (p. 224) as 'a gift of reverent and grateful love' from Bely to his 'faithful guide, teacher and inspiration'.

46. D. S. Mirsky, *Contemporary Russian Literature, 1881–1925* (Knopf, New York, 1926), pp. 181–2.

47. Mochulsky, *Andrey Bely*, p. 50.

48. Mirsky, *Contemporary Russian Literature*, p. 225.

49. Mochulsky, *Andrey Bely*, p. 147.

50. M. Slonim, *From Chekhov to the Revolution* (Oxford University Press, New York, 1962), pp. 192–3.

51. See V. Ivanov's article, 'The Inspiration of Horror', discussed in Mochulsky, *Andrey Bely*, pp. 154–5.

52. A. Bely, *Petersburg* (Grove Press, New York, 1959, translated by J. Cournos), p. xii.

53. Ibid., p. 115.

54. Ibid., p. 234.

55. Mochulsky, *Andrey Bely*, p. 152. Bely himself makes this connection explicit in the particular episode in question writing of Dudkin as 'the new Evgeny' and noting 'Evgeny's fate was about to be repeated' before the

astonishing entrance of the Bronze Horseman. The following passage is
significant (p. 234):

> The bronze giant had been galloping through ages of time and, reaching the
> present moment, had completed a cycle; ages had sped by; Nicholas I had
> ascended to the throne; and, after him, the Alexanders; and Alexander
> Ivanovich Dudkin, himself a shadow, had restlessly overcome the ages day
> after day, year after year, roaming up and down the Petersburg prospects –
> awake and in dream. The clanging thunder of metal had pursued him and all
> the others, shattering their life.

56. Pushkin, *The Bronze Horseman*, lines 84–5.

57. J. D. Elseworth, *Andrey Bely* (Bradda Books, Letchworth, 1972), p. 80.

58. R. Ivanov-Razumnik in Field, *Complection*, p. 190. The reference here to
the 'Dostoevskian motif' probably refers to the following passage in *A Raw
Youth*:

> A hundred times over is such a fog, I have been haunted by a strange and
> persistent fancy: 'what if this fog should part and float away? Would not all
> this rotten and slimy town go with it, rise up with the fog, and vanish like
> smoke, and the old Finnish marsh be left as before and in the midst of it,
> perhaps to complete the picture, a bronze horseman on a panting, over-driven
> steed?'

See also the discussion of the famous 'Vision of the Neva' written by Dostoevsky
in 1861 as a retrospective accounting of the passage included in his early short
story, 'A Weak Heart' (1848); in J. Frank, *Dostoevsky: the Seeds of Revolt
1821–1849* (Princeton University Press, Princeton, NJ, 1976), pp. 133–6 and
318–22.

59. Renato Poggioli, 'Kafka and Dostoevsky' in Angle Flores (ed.), *The
Kafka Problem* (Octagon Books, New York, 1963), p. 99.

60. See, for example, S. Gibson, 'Sense of Place and Defense of Place: a Case
Study of the Toronto Islands', unpublished PhD dissertation, University of
Toronto, 1980.

61. Fanger, *Creation of Nikolai Gogol*, p. viii.

12 GEORGE CRABBE'S SUFFOLK SCENES

Hugh C. Prince

In a history of English literature, George Crabbe (1754–1832) stands apart from his contemporaries.[1] He was an Augustan poet who rhymed couplets in the manner of Pope, Gray and Dyer, but his verses destroyed the pastoral idyll and depicted village life, 'as Truth will paint it, and as Bards will not'.[2] He rejected Thomson's progressive view of the seasonal round and Goldsmith's nostalgic vision of the deserted village. The countryside which Crabbe knew most intimately was not an Elysium where swains and shepherdesses idly disported themselves, but a stretch of 'burning sand' where men and women struggled ceaselessly to wrest a meagre subsistence from the soil. The land was cultivated by dint of back-breaking toil, the harvest was paid for in sweat, and work continued in fair weather and foul. Crabbe spared no harrowing detail in exposing the sufferings of 'the poor laborious natives of the place' and he explicitly connected the degradation of the labourer with exploitation by farmers and landowners:

> Where Plenty smiles — alas! she smiles for few —
> And those who taste not, yet behold her store,
> Are as the slaves that dig the golden ore, —
> The wealth around them makes them doubly poor.[3]

Among oppressors of the poor, Crabbe counted his own readers, to whom he appealed:

> Nor mock the misery of a stinted meal;
> Homely, not wholesome, plain, not plenteous, such
> As you who praise would never deign to touch.[4]

In reproaching 'gentle souls who dream of rural ease', he attacked not only his patrons, but also exposed his own sheltered position as a country parson. He challenged middle-class consciences by taunting his readers:

> How would ye bear in real pain to lie,
> Despised, neglected, left alone to die?[5]

Crabbe, the Augustan rebel, gained no support for his protests from the Romantic poets. Clinging tenaciously to a hard life on hungry soils

in east Suffolk, he did not share the exalted enthusiasm of the Lake
Poets for bare rocks, tumbling waterfalls and wild fells. Coleridge
dismissed Crabbe for 'an absolute defect of the high imagination', and
Hazlitt was repelled by Crabbe's insistent realism and tedious recital of
disgusting details such as slime, tar, weeds and bilge-water.[6] Romantic
poetry tried to elevate the spirit and invest commonplace objects with
eternal, universal meaning, whereas Crabbe narrated 'homespun griefs
in homespun verse', describing incidents plainly, 'without an
atmosphere'.[7]

Crabbe's work probably had little influence on the development of
English literature, but it is of great interest in expressing the regional
distinctiveness of the Sandlings, in evoking a sense of place, and in
projecting emotional experiences into landscape symbols. I shall
examine Crabbe's descriptions of east Suffolk in the light of his
personal associations. I shall follow his search for the reality of the
Sandlings, comparing his views with those of agricultural writers. And I
shall review his explorations of scenes in the minds of his characters.

Possessed by his Native Heath

Crabbe's love -hate relationship with Aldeburgh lasted from the day he
was born, Christmas Eve 1754, to the day he died, 3 February 1832.
The roots of his attachment to his birthplace go back through his
father, also named George Crabbe, to his grandfather, yet another
George Crabbe, and spread through numerous branches of a large
family who farmed in the neighbourhood or went to sea.[8] The poet's
father kept a warehouse on Slaughden Quay and held office as Salt-
master, collector of salt duties in Aldeburgh. The poet's mother was a
publican's widow. The poet was the oldest of six children, of whom
four sons and a daughter survived to adulthood. Of the younger
children, Robert became a glazier, John became captain of a Liverpool
slave-ship, William, taken prisoner at sea by Spaniards, reached Mexico
where he prospered as a silversmith and there raised a large family, and
Mary married a builder in Aldeburgh. The emotional shock caused by
the death of another sister in infancy haunted the poet for the rest of
his life:

> For then first met and moved my early fears,
> A father's terrors, and a mother's tears.
> Though greater anguish I have since endured, —

> Some heal'd in part, some never to be cured;
> Yet was there something in that first-born ill,
> So new, so strange, that memory feels it still.[9]

It was his first experience of grief, tinged by a sense of discord within the home. Crabbe's mother was loving, patient and deeply religious. Crabbe's father was stern, violent when he drank, skilled in business and exceptionally talented in calculation. When the children were young he was a proud and attentive parent, but as they grew older he became quarrelsome and negligent. On the other hand, he encouraged the poet to pursue his studies. Crabbe recalled an occasion when, as a small child, he accompanied his father on a yacht club outing. Upon reaching a neighbouring town at mid-day:

> The men drank much, to whet the appetite;
> And, growing heavy, drank to make them light;
> Then drank to relish joy, then further to excite.
> Their cheerfulness did but a moment last;
> Something fell short, or something overpast.
> The lads play'd idly with the helm and oar,
> And nervous women would be set on shore,
> Till 'civil dudgeon' grew, and peace would smile no more.[10]

From an early age, Crabbe felt awkward handling a boat, ill at ease among boys of his own age, not fully accepted in the company of men at work, and regarded with hostility by people on the shore. He repeatedly described people he met in his own home town as amphibians and strangers:

> I often rambled to the noisy quay,
> Strange sounds to hear, and business strange to me;
> Seamen and carmen, and I know not who,
> A lewd, amphibious, rude, contentious crew.[11]

He kept his distance from the shore folk:

> Here joyless roam a wild amphibious race,
> With sullen woe display'd in every face;
> Who, far from civil arts and social fly,
> And scowl at strangers with suspicious eye.[12]

On solitary walks across heaths and marshes, he collected plants, watched birds, met lonely shepherds, listened to fishermen and old widows. He was a studious boy, reading widely and learning avidly.

From the age of seven he went away to school, first at Bungay, later at
Stowmarket. In 1771 he was apprenticed to a doctor in Woodbridge
but received little formal training until, five years later, as a last resort,
he decided to attend lectures in London. By this time, his life had
changed fundamentally. He had fallen in love with Sarah Elmy, niece of
the squire of Parham Hall, but he had no means of supporting her. She
appreciated his poetry and inspired him to write. As a doctor, he was
unable to earn sufficient even to feed himself and was obliged to take
casual employment in his father's warehouse.

Living at home in his twenties, Crabbe grew more and more discon-
tented and restless. He felt confined in the midst of all that empty
space:

> Where all beside is pebbly length of shore,
> And as far as eye can reach, it can discern no more.[13]

The eternal sameness of the shingle numbed the senses but the boundless sea, the open sky and the distant horizon beckoned a roving spirit.
The sea was finely attuned to Crabbe's changing moods: his anger burst
into storm, his boredom settled into a monotonous rhythm. On a calm
summer day, he gazed disconsolately at the ocean:

> . . . swelling as it sleeps,
> Then slowly sinking; curling to the strand,
> Faint, lazy waves o'ercreep the ridgy sand,
> Or tap the tarry boat with gentle blow,
> And back return in silence, smooth and slow.[14]

The gentle beat of waves on a wooden hull and the faint aroma of warm
tar would be perceptible only when everything else was still. The
images of a coastline receding into the distance and a calm sea convey a
sense of emptiness, inactivity and desolation. Looking back over the
Alde estuary, the oppressive weight of drabness and endless repetition
become almost unbearable:

> At the same time the same dull views to see,
> The bounding marsh-bank and the blighted tree;
> The water only, when the tides were high,
> When low, the mud half-cover'd and half-dry;
> The sun-burnt tar that blisters on the planks,
> And bank-side stakes in their uneven ranks;
> Heaps of entangled weeds that slowly float,
> As the tide rolls by the impeded boat.[15]

The ceaseless ebb and flow of the tide, lapping monotonously against another moored boat, accompanied by another faint whiff of warm tar, echo the sounds and smells of the shingle beach. On the bed of the Alde at low water, expanses of slimy mud glisten in the sun:

> Here dull and hopeless he'd lie down and trace
> How sidelong crabs had scrawl'd their crooked race;
> Or sadly listen to the tuneless cry
> Of fishing gull or clanging golden-eye;
> What time the sea-birds to the marsh would come,
> And the loud bittern, from the bullrush home,
> Gave from the salt-ditch side the bellowing boom:
> He nursed the feelings these dull scenes produce.[16]

It is a deeply introspective view of landscape.

Crabbe felt himself slowly slipping and sinking into this dull mud. His misery and his increasing aggression towards his father were alluded to in 'Peter Grimes'. Crabbe's father now drank heavily and quarrelled violently with his grown-up son, who not only seemed incapable of earning a living in the profession for which he had been educated, but worked sullenly and grudgingly in the warehouse, planned to marry a genteel girl and spent much time botanising and writing poetry. Crabbe's mother, suffering from dropsy, grew weaker, but her son still relied on his

> . . . mother's kindness, and the joy
> She felt in meeting her rebellious boy.[17]

Crabbe's feelings of resentment towards his father found expression in 'The Village', an account of a winter storm that struck eleven houses with ungovernable paternal fury. Crabbe watched helplessly as the waves battered down roofs, tore away walls and swept the debris down the beach. Again and again, he cried out pityingly for those 'who still remain':

> . . . Ah! hapless they who still remain;
> Who still remain to hear the ocean roar,
> Whose greedy waves devour the lessening shore;
> Till some fierce tide, with more imperious sway,
> Sweeps the low hut and all it holds away.[18]

Shortly after that storm, Crabbe resolved to leave Aldeburgh and seek his fortune as a writer:

> So waited I the favouring hour, and fled;
> Fled from these shores where guilt and famine reign.[19]

Early in 1780, with his best verses and three pounds in his pocket, he sailed for London. His arrival in the capital was not greeted with immediate success. He failed to find work as a writer. Booksellers and rich patrons rejected his manuscripts. By the end of the year he was penniless and hungry. In desperation, he wrote to Edmund Burke, begging for support. On the following day, he was granted an interview and the conservative philosopher not only showed an interest in 'The Library' and 'The Village' but offered the poor poet patronage. Burke secured the publication of the two works, helped Crabbe to enter the church, helped him to obtain a curacy at Aldeburgh, and later introduced him to the Duke of Rutland by whom he was appointed chaplain at Belvoir Castle.

In 1783 Crabbe married Sarah Elmy and in 1787 became rector of Muston in the vale of Belvoir and of Allingham in Lincolnshire. With his wife and growing family he made frequent excursions into Suffolk. He went back to the Sandlings to renew old acquaintances and to revisit scenes of his childhood. In 1792, on the death of her uncle, his wife inherited Parham Hall and adjoining farms. Crabbe left a curate at Muston and moved to Suffolk, first to Parham, then to Glemham Hall, lastly to a house in Rendham. He assisted local parsons, returning after a time to Muston to resume his pastoral duties, but he spent more time reading, writing and studying natural history. In Suffolk he was often fretful and dissatisfied, but after a long period away from the district he would be overcome by nostalgia. He would ride over a hundred miles to glimpse the sea and his native heath:

> Strange, that a boy could love these scenes, and cry
> In very pity -- but that boy was I.[20]

In 1813, following his wife's death, he took up the living of Trowbridge in Wiltshire and remained there until he died, at the age of seventy-eight, in 1832. While resident in Wiltshire, he continued to write about the Sandlings but never turned his pen to West Country scenes.

The Reality of the Sandlings

Crabbe's east Suffolk countryside was not a wonder of nature to be contemplated from a distance, a sublime vision to be recollected in tranquillity. It was not sacred ground to be protected from rash assault or defiling hands, to be worshipped in a romantic trance. Nor was it a pretty landscape, to be looked at, dressed up and rhapsodised with pastoral sentiments. Crabbe was neither a romantic nor a pastoralist. He was a naturalist and a stern realist. He studied plants, birds and muddy estuaries, not far-off mountains and vast solitudes. Country life was not genial and peaceful, but harsh and exhausting.

Crabbe scorned artifice and the pretentious mannerisms of the picturesque cult. He despised ostentation in landscape gardening and contrasted old-fashioned wisdom with modern follies:

> I miss the grandeur of the rich old scene,
> And see not what these clumps and patches mean.[21]

He was sad to see newcomers altering venerable old houses and gardens and he lamented the loss of precious associations. Of a new park, he wrote:

> The things themselves are pleasant to behold,
> But not like those which we beheld of old, —
> That half-hid mansion, with its wide domain,
> Unbound and unsubdued! — but sighs are vain.[22]

Above other old things, 'he in reverence held the living wood, that widely spreads in earth the deepening root'.[23] He condemned the vanity of attempting to improve upon nature and he objected to particular means employed in achieving these effects. He objected to cutting down growing trees. He objected to wasting cultivable land. He objected to men's labour being misused. He reproved a landowner,

> Who, while unvalued acres ran to waste,
> Made spacious rooms, whence he could look about,
> And mark improvements as they rose without:
> He fill'd the moat, he took the wall away,
> He thinn'd the park, and bade the view be gay.[24]

Crabbe was carried away by the force of his indignation and, rather implausibly, inflicted financial ruin upon this extravagant squire:

> The scene was rich, but he who should behold
> Its worth was poor, and so the whole was sold.[25]

Poetically, Crabbe punished the squire not for bad taste but for improvidence. The squandering of funds and the failure to use land profitably he judged more severely than the offence of defacing natural beauty. Crabbe was a somewhat puritanical utilitarian rather than an aesthete.

He appraised land with an agricultural writer's eye. He shared a farmer's pride in 'that unrivall'd flock, . . . the village boast, the dealer's theme'.[26] Appreciating a good harvest successfully gathered, he enjoyed walking over 'stubble ground, where late abundance stood'.[27] But as he walked, he was made increasingly uneasy by reflecting that the improver, whilst proud of his bountiful crop, had done little to earn the affections of his labourers: 'we our poor employ, and much command, though little we enjoy'.[28] For a few retired bankers and merchants, country life was an amusing distraction from the cares of the city and afforded some consolation for the disappointments accompanying old age:

> We plant a desert, or we drain a fen;
> And – here, behold my medal! – this will show
> What men may merit when they nothing know.[29]

Crabbe could not conceal his contempt for gentlemen who played at farming for a recreation. Without other sources of income, full-time farmers on the lighter soils could expect to gain no more than a bare subsistence, and many failed.

The Sandlings contain numerous remains of abandoned fields and farms. Surveying a tract of bog and heath, he observed:

> Here pits of crag, with spongy, plashy base,
> To some enrich th'uncultivated space:
> For there are blossoms rare, and curious rush,
> The gale's rich balm, and sun-dew's crimson blush
> Whose velvet leaf with radiant beauty dress'd,
> Forms a gay pillow for the plover's breast.[30]

The soils here, too sandy and dry to raise a crop, proved too infertile even to be redeemed by the addition of the underlying crag. Crabbe perceived the irony of rare wildlife taking refuge in pits dug to reclaim the heaths. Arthur Young reported that 'the understratum of this district varies considerably, but in general it may be considered as sand, chalk, or *crag*; in some parts marl or loam. The crag is a singular body of cockle and other shells, found in great masses in various parts of the country from Dunwich quite to the river Orwell.'[31] Young noted that pits were dug from fifteen to twenty feet into the crag 'for improving the heaths', but ignored traces of deserted fields where the attempt had failed.

Crabbe, the plant collector, was painfully reminded that success in heath reclamation was achieved not only through arduous labour but by ruthlessly eradicating weeds. The creator of the 'model for all farms' boasted:

> 'Look at that land – you find not there a weed,
> We grub the roots and suffer none to seed,
> To land like this no botanist will come.'[32]

Like most countrymen, Crabbe also drew a sharp distinction between useful and harmful plants and understood clearly the dire effects of infestation:

> Rank weeds, that every art and care defy,
> Reign o'er the land, and rob the blighted rye:
> There thistles stretch their prickly arms afar,
> And to the ragged infant threaten war;
> There poppies nodding, mock the hope of toil;
> There the blue bugloss paints the sterile soil;
> Hardy and high, above the slender sheaf,
> The slimy mallow waves her silky leaf;
> O'er the young shoot the charlock throws a shade,
> And clasping tares cling round the sickly blade.[33]

Paradoxically, the hungry sand supports a luxuriant growth of weeds, while starving and choking cultivated crops.

Crabbe took a gloomier view of the agricultural value of the Sandlings than either John Kirby who, in 1764, saw promise in many hundreds of acres of heaths and sheep-walks already 'converted into good Arable Land', or Arthur Young who, in 1804, praised the district as 'one of the best cultivated in England; not exempt from faults and deficiencies, but having many features of unquestionably good management . . . there are few districts in the county abounding with wealthier farmers'.[34] Further south, approaching Ipswich in March 1830, William Cobbett reported the loam soils

in such a beautiful state, the farmhouses all white, and all so much alike; the barns and everything about the homesteads so snug; the stocks of turnips so abundant everywhere; the sheep and cattle in such fine order; the wheat all drilled; the ploughman so expert; the furrows, if a quarter of a mile long, as straight as a line, and laid as truly as if with a level; in short, here is everything to delight the eye, and to make the people proud of their country.[35]

This peculiarly flattering picture may have been coloured by Cobbett's prejudice against Suffolk as

a very *highly-favoured county*: it has had poured into it millions upon millions of money, drawn from Wiltshire and other inland counties. I should suppose that Wiltshire alone has, within the last forty years, had two or three millions of money drawn from it, *to be given to Essex and Suffolk*. At one time there were not less than sixty thousand men kept on foot in these counties. The increase of London, too, the swelling of the immortal Wen, have assisted to heap wealth upon these counties; but, in spite of all this, the distress pervades all ranks and degrees, except those who live on the taxes.[36]

In Crabbe's view, neither the stationing of troops at Aldeburgh nor the purchase of estates by fundholders in any way alleviated the abiding poverty of his native heath.

Writing sixteen years after Crabbe's death, Hugh Raynbird indicated differences within the Sandlings, between undrained marshes, productive loams and sterile sands. He reported that the district 'is more or less of a sandy nature, a great portion of which is highly cultivated, though in some parts the soil is of a very inferior description, sometimes a blowing sand, and still lying almost waste'. Riding over long stretches of wild heath, Raynbird was impressed by the widespread 'appearance of barrenness' and the crops he saw in the fields gave 'ample evidence of the sterility of the soil'. Only between Sutton and Woodbridge did a tract of heathland appear, 'from the luxuriant growth of the ferns and whins, to be of a better quality'.[37] Amid the bracken, the broom and the leggy heather, Crabbe was at home.

Emotional Environments

In Crabbe's poetry, his own emotions and those of his characters are reflected in their environments. Feelings of joy, outbursts of anger and changes of mood are symbolised by objects in environments and are transmuted through landscape changes.

> It is the Soul that sees; the outward eyes
> Present the object, but the Mind descries;
> And thence delight, disgust, or cool indiff'rence rise.[38]

Earlier poets described how beautiful scenes aroused pleasure, how smoothness soothed and sunlight warmed the spirit, how the sublimity of mountains and storms inspired awe and struck terror into travellers' minds, how picturesque landscapes evoked painterly feelings. Crabbe not only recounted how a landscape affected an observer, but he indicated how an observer projected his inner mood onto his environment:

> When minds are joyful, then we look around,
> And what is seen is all on fairy ground;
> Again they sicken, and on every view
> Cast their own dull and melancholy hue.[39]

In 'The Lover's Journey', on a June morning, Orlando rides out of Aldeburgh across the Sandlings into the woodland district of mid-Suffolk to meet his Laura. His heart is filled with fond anticipation; the barren heaths echo his joy:

> 'This neat low gorse,' said he, 'with golden bloom,
> Delights each sense, is beauty, is perfume;
> And this gay ling, with all its purple flowers,
> A man at leisure might admire for hours;
> This green-fringed cup-moss has a scarlet tip,
> That yields to nothing but my Laura's lip;
> And then how fine this herbage! men may say
> A heath is barren; nothing is so gay.
> Barren or bare to call such charming scene
> Argues a mind possess'd by care and spleen.'[40]

He passes over bogs, marshes, fens and wild commons, exulting at each delightful prospect, enchanted by the rich profusion of colours.

On reaching his destination, he is handed a note telling him that Laura has gone away to see a friend. He is desolated, but decides to follow her to her friend's house. Through mid-Suffolk's verdant pasture

country, he rides along lanes overhung by tall timbered hedgerows, by
the edge of a limpid stream, glimpsing elegant mansions in spacious
grounds:

> 'I hate these scenes,' Orlando angry cried,
> 'And these proud farmers! yes, I hate their pride:
> See! that sleek fellow, how he strides along,
> Strong as an ox, and ignorant as strong;
> Can yon close crops a single eye detain
> But he who counts the profits of the grain?
> And these vile beans with deleterious smell,
> Where is their beauty? can a mortal tell?'[41]

The stench of rotting beans was Crabbe's pet aversion. More than
chance directed his footsteps, when he felt miserable, towards a bean
field at the very season when the beans were decaying. The environment
was contrived to respond to the observer's mood.

Orlando catches up with Laura and the lovers are reconciled. On the
morrow, blissfully content, they return home together by the way they
came. First, through mid-Suffolk,

> By the rich meadows where the oxen fed,
> Through the green vale that form'd the river's bed;
> And by unnumber'd cottages and farms,
> That have for musing minds unnumber'd charms.[42]

After leaving Laura, Orlando continues his journey alone, back across
the marshes and barren sands of east Suffolk.

Totally absorbed in his private thoughts, he is oblivious to his
surroundings:

> Then could these scenes the former joys renew?
> Or was there now dejection in the view? –
> Nor one or other would they yield – and why?
> The mind was absent, and the vacant eye
> Wander'd o'er viewless scenes, that but appear'd to die.[43]

Crabbe rings the changes of mood and activity from elation to dejection,
from restlessness to daydreaming, and describes these feelings not by
registering changes in facial expressions or tones of voice but by looking
through his characters' eyes at different aspects of the environment.

A most remarkable literary feat is the rendering of an outlook of blank despair. Crabbe observed that in a state of depression the mind is unresponsive to external environments and is aroused only by pain:

> Oft when the feet are pacing o'er the green
> The mind is gone where never grass was seen,
> And never thinks of hill, or vale, or plain,
> Till want of rest creates a sense of pain,
> That calls that wandering mind, and brings it home again.[44]

A visit to a condemned prisoner in a cell at Newgate Prison left an indelible impression on the young Crabbe. He pondered on what he had seen and analysed his feelings. His report is a dry, clinical assessment:

> Since his dread sentence, nothing seem'd to be
> As once it was — he seeing could not see,
> Nor hearing, hear aright; — when first I came
> Within his view, I fancied there was shame,
> I judged resentment; I mistook the air, —
> These fainter passions live not with despair;
> Or but exist and die: — Hope, fear, and love,
> Joy, doubt, and hate may other spirits move,
> But touch not his, who every waking hour
> Has one fix'd dread, and always feels its power.[45]

Terror drives out every other sensation and blots out sights, sounds, smells and textures from the environment. Unlike Edmund Burke's spectator in high mountains who was awed by the sublimity of his surroundings, Crabbe's prisoner was petrified from within by his own dreadful apprehension. The prisoner was addressed in the third person and given no name. Crabbe imagined that in a state of mortal terror the mind would withdraw into a dreamworld and would create intensely felt private images of environments. He conjured for the condemned man a nightmare spectacle of his own execution:

> There crowds go with him, follow and precede;
> Some heartless shout, some pity, all condemn,
> While he in fancied envy looks at them;
> He seems the place for that sad act to see,
> And dreams the very thirst which then will be.[46]

This awful, searing apparition is followed by some of the most
radiantly beautiful lines Crabbe composed, recalling a summer day
some years earlier when the prisoner, with his lover,

> Stray o'er the heath in all its purple bloom, --
> And pluck the blossom where the wild bees hum;
> Then through the broomy bound with ease they pass,
> And press the sandy sheep-walk's slender grass,
> Where dwarfish flowers among the gorse are spread,
> And the lamb browses by the linnet's bed;
> Then 'cross the bounding brook they make their way
> O'er its rough bridge -- and there behold the bay! --
> The ocean smiling to the fervid sun -
> The waves that faintly fall and slowly run.[47]

This sweet, peaceful stroll, passing through all the spots closest to
Crabbe's heart, is ended abruptly:

> She cries: -- Alas! the watchman on his way
> Calls, and lets in -- truth, terror, and the day![48]

The irony of the poem is that Crabbe's most purely pleasurable vision
of the Sandlings appears in a dream given to a man totally cut off from
reality.

Crabbe returns again and again to moments of sadness and regret. In
'Delay has Danger', Henry, betraying his true love, broods over his
faithlessness. At sunrise -- normally an occasion for hopeful thoughts --
he goes down to the shore, chilled by a fierce wind. The wind churns up
the water, tossing and bending dark pines against a reddening sky:

> Before him swallows, gathering for the sea,
> Took their short flights, and twitter'd on the lea;
> And near the bean-sheaf stood, the harvest done,
> And slowly blacken'd in the sickly sun;
> All these were sad in nature, or they took
> Sadness from him, the likeness of his look,
> And of his mind -- he ponder'd for a while,
> Then met his Fanny with a borrow'd smile.[49]

Henry's agitation and his grief are exactly mirrored in the features of
the cold turbulent dawn. In another waterside scene, drawn from the

poet's childhood memories, the young Crabbe looks forward to a pleasant day on the river:

> Sweet was the morning's breath, the inland tide,
> And our boat gliding, where alone could glide
> Small craft — and they oft touch'd on either side.
> It was my first-born joy.[50]

The day ended with the adults quarrelling drunkenly and the child sadly disappointed. His change of mood is symbolised by the onset of a rainstorm:

> Now on the colder water faintly shone
> The sloping light — the cheerful day was gone;
> Frown'd every cloud, and from the gathered frown
> The thunder burst, and rain came pattering down.
> Now, all the freshness of the morning fled,
> My spirits burden'd, and my heart was dead.[51]

The promised joy on setting out deepened the misery suffered on returning. The child's excursion provided a foretaste of the lover's journey. Crabbe traced the swing from hope to despair through changes in weather, lighting, sounds, smells, textures and associations. He identified emotional changes with changes in the *feel* of environments.

Water is the most portentous of all landscape elements. Its changing moods fascinated Crabbe. In 'The Parting Hour', childhood sweethearts can see no prospect of affording a home when, unexpectedly, the boy receives an invitation to go to the West Indies to make his fortune. They part, significantly, at the water's edge. The ship in which he sails away is captured by Spaniards and, like Crabbe's younger brother William, he never reaches the West Indies. Many years later, as an old man who has lost a fortune, he returns to his sweetheart. Crabbe expresses foreboding at the moment of departure in a marine analogy of powerful simplicity:

> So while the waters rise, the children tread
> On the broad estuary's sandy bed;
> But soon the channel fills, from side to side
> Comes danger rolling with the deep'ning tide;
> Yet none who saw the rapid current flow
> Could the first instant of that danger know.[52]

Crabbe's most persistent mood was fatalistic and most of his environments were laden with doom, not melodramatically but in the simple, matter-of-fact way that the tide comes and goes. Dark, cold water reflected his deep melancholy, warm water excited his anger, while waves and ripples moved in time with his agitation.

Crabbe's Legacy to the Sandlings and to Sensibility

Stratford-upon-Avon is known throughout the world as Shakespeare's birthplace, but Shakespeare rose to fame in London and his work belongs to mankind everywhere. Aldeburgh, on the other hand, is not merely Crabbe's birthplace, it is immortalised in his verses, and he made the place his very own. The Aldeburgh Festival celebrates an extensive repertoire of music by Benjamin Britten, but it owes its special character to the original performance of 'Peter Grimes'.[53] Aldeburgh can no more exist without Crabbe than it can ignore the encroaching sea or transform the Alde estuary into a Norwegian fiord. Crabbe belonged to Aldeburgh yet had the urge to escape from it; he also learned to stand back and reflect upon it. His language and imagery breathe the genius of the place. His words are inscribed on its streets, shingle, mud, boats and blistering tar. Crabbe depicted these familiar scenes with the accuracy of a Dutch master: 'he paints in words instead of colours', wrote Hazlitt, 'there is no other difference'.[54] Byron admired the pictorial quality in Crabbe's writing, describing him as 'Nature's sternest painter, yet the best'.[55] Like his Suffolk contemporary John Constable, he devoted painterly attention to local details, to subtleties of colour, to effects of light and shade and to the changing moods of sky and sea.

Crabbe belongs to Aldeburgh, but most Aldeburgh people would prefer to disown him. Few of his fellow townsmen would consult him as patients or attend his church services and his poetical works won him few friends in the town during his lifetime or since. His stark realism was too disturbing for popular comfort. Like other writers associated with other towns -- Arnold Bennett with the Square in Burslem, D. H. Lawrence with the mining village of Eastwood in Nottinghamshire, Sinclair Lewis with Main Street in Sauk Center, Minnesota, and John Steinbeck with Cannery Row in Monterey, California -- Crabbe presented too cruel a picture to mollify local opinion. A Suffolk historian, W. G. Arnott, complains that Crabbe depresses him: 'He dug deep into human nature and seemed to seek

want, woe and wretchedness in all his writings.'[56] Ronald Blythe's
biographical portrait of the nearby village of Akenfield makes no
reference to Crabbe's tales from Aldeburgh, and Hugh Raynbird, whose
assessment of the Sandlings accords closely with Crabbe's, hails an
inferior Robert Bloomfield as 'the Suffolk poet'.[57] Like many before
him, Crabbe remains a prophet without glory in his own country.

Those who share the views of residents in condemning him for
painting too gloomy a picture and those who praise him for presenting
nothing but the plain unvarnished truth will join in acclaiming him as
the originator of the Suffolk Sandlings or, at least, the one above all
others to establish the image of the region. Crabbe's image of heath and
marsh, muddy estuary and shingle ridge, has been adopted, either
reluctantly or enthusiastically, by guidebook writers, landowners, forest
officers, county planners and building developers. Obsolete warehouses
and fishermen's huts are carefully restored, particularly by newcomers.
Bird sanctuaries and remnants of heath are protected and no change in
the appearance of Crabbe's favourite scenes passes without comment.
Extensive changes have been made in the Sandlings since Crabbe lived,
but openness — the characteristic he described most sensitively —
remains dominant in the landscape.

Crabbe's poems did not arouse feelings of reverence for nature as
passionate as those aroused by Wordsworth, nor were consciences
stirred against poverty and social injustices as deeply by Crabbe as by
Clare, but Crabbe's obsession with the regional distinctiveness of the
Sandlings inspired both Walter Scott and Thomas Hardy to search out
the identities of regions and localities in Scotland and in southern
England. Scott and Hardy were among Crabbe's most devoted readers
and they described places in a similarly realistic manner. And just as
features of Crabbe's Sandlings have been preserved and imitatively
revived, so landscapes in the Scottish Highlands and in Wessex have
been redesigned or reconstructed to look like scenes depicted by Scott
or Hardy.

Crabbe invested low shorelines and coastal marshes with menace and
mystery, tones deepened and thickened in Dickens' descriptions of the
north Kent marshes in *Great Expectations* and of the beach at
Lowestoft in *David Copperfield*. Philip Larkin's descriptions of the
Humber marshes recall Crabbe's sadness and feeling of emptiness.
Tennyson's 'Northern Farmer' takes up Crabbe's refrain on the bareness
and variety of wildlife on the Lincolnshire heath. Thomas Hardy
amplifies Crabbe's protest at the futility of attempting to cultivate
barren sands. George Eliot, like Crabbe, regards water as an agent of

death. Crabbe brought to landscape description a new sensibility. He related characters to their environments, indicating how places affected people and how individuals' moods were echoed and expressed in the scenes around them.

Notes

I am grateful to my father, Louis Prince, for passing on his artistic appreciation of Suffolk, and to my colleagues at University College, London, particularly Jacquelin Burgess, Stephen Daniels, David Lowenthal, Catherine Middleton and John Thornes for their valuable comments on an early draft of this chapter.

1. My guides to the critical literature have been Lilian Haddakin, *The Poetry of Crabbe* (Chatto & Windus, London, 1955); F. R. Leavis, *Revaluation: Tradition and Development in English Poetry* (Chatto & Windus, London, 1936), pp. 124–9; F. L. Lucas, *George Crabbe: an Anthology* (Cambridge University Press, Cambridge, 1933); Howard Mills, *George Crabbe: Tales, 1812 and other Selected Poems* (Cambridge University Press, Cambridge, 1967); Peter New, *George Crabbe's Poetry* (Macmillan, London, 1976); Arthur Pollard, *Crabbe: the Critical Heritage* (Routledge & Kegan Paul, London, 1972); Raymond Williams, *The Country and the City* (Chatto & Windus, London, 1973), pp. 13–95.

2. Later editors have made minor changes in spelling and punctuation. I have quoted from the first posthumous edition, edited by George Crabbe (the poet's son), *The Poetical Works of the Rev. George Crabbe, LL.B: with his Letters and Journals, and his Life* (8 vols., John Murray, London, 1834), hereafter referred to as *Works*. The reference here is to 'The Village', *Works*, II, p. 76.

3. Ibid., p. 79.

4. Ibid., pp. 80–1.

5. Ibid., p. 84.

6. S. T. Coleridge, *Table Talk and Omniana*, ed. T. Ashe (1884), p. 276, cited in New, *George Crabbe's Poetry*, p. 4; William Hazlitt, *The Spirit of the Age*, 1825 (World's Classics, Oxford University Press, London, 1947), p. 248.

7. Leslie Stephen, *Hours in a Library*, 1876, cited in Haddakin, *Poetry of Crabbe*, p. 11; *Quarterly Review*, vol. IV (1810), pp. 281–312, reprinted in Pollard, *Critical Heritage*, p. 125.

8. I followed the *Life* written by the poet's son, being vol. I of the *Works*.

9. 'Infancy', *Works*, IV, p. 103.

10. Ibid., p. 105.

11. 'The Adventures of Richard', *Works*, VI, p. 85.

12. 'The Village', *Works*, II, p. 77.

13. 'The Borough', *Works*, III, p. 25.

14. Ibid.

15. 'Peter Grimes', *Works*, IV, pp. 45–6.

16. Ibid., p. 46.

17. 'The Adventures of Richard', *Works*, VI, p. 85.

18. 'The Village', *Works*, II, p. 79.

19. Ibid.

20. 'The Adventures of Richard', *Works*, VI, p. 85.

21. 'The Ancient Mansion', *Works*, VIII, p. 164.

22. Ibid.

23. 'The Hall', *Works*, VI, p. 20.

24. Ibid., p. 21.
25. Ibid.
26. 'The Adventures of Richard', *Works*, VI, p. 73.
27. Ibid.
28. Ibid., p. 74.
29. Ibid.
30. 'The Borough', *Works*, III, p. 23.
31. Arthur Young, *General View of the Agriculture of the County of Suffolk*, 3rd edn (Richard Phillips, London, 1804), p. 5.
32. 'The Adventures of Richard', *Works*, VI, pp. 72–3.
33. 'The Village', *Works*, II, p. 77.
34. John Kirby, *The Suffolk Traveller*, 2nd edn (Longman, London, 1764), p. 2; Young, *General View*, p. 5.
35. William Cobbett, *Rural Rides*, 1830 (2 vols., Dent, London, 1948, Everyman edition), vol. 2, p. 225.
36. Ibid., vol. 2, p. 226.
37. Hugh Raynbird, 'On the Farming of Suffolk', *Journal of the Royal Agricultural Society of England*, vol. 8 (1848), pp. 263–4.
38. 'The Lover's Journey', *Works*, V, p. 21.
39. Ibid.
40. Ibid., pp. 22–3.
41. Ibid., pp. 30–1.
42. Ibid., p. 34.
43. Ibid.
44. 'The Adventures of Richard', *Works*, VI, p. 75.
45. 'Prisons', *Works*, IV, pp. 67–8.
46. Ibid., p. 69.
47. Ibid., pp. 70–1.
48. Ibid., p. 71.
49. 'Delay has Danger', *Works*, VII, p. 70.
50. 'Infancy', *Works*, IV, p. 104.
51. Ibid., p. 105.
52. 'The Parting Hour', *Works*, IV, p. 178.
53. E. M. Forster, *Two Cheers for Democracy*, 1951 (Edward Arnold, London, 1972), 'George Crabbe and Peter Grimes', pp. 166–80.
54. 'The Pictorial Element' in Haddakin, *The Poetry of Crabbe*, p. 128.
55. Byron, 'English Bards and Scotch Reviewers' in H. Milford (ed.), *The Poetical Works of Lord Byron* (Oxford University Press, London, 1949), p 121.
56. W. G. Arnott, *Alde Estuary: the Story of a Suffolk River* (Norman Adlard, Ipswich, 1952), p. 86; more sympathetic are Julian Tennyson, *Suffolk Scene* (Blackie & Son, London, 1939); and Norman Scarfe, *The Suffolk Landscape* (Hodder and Stoughton, London, 1972), p. 27.
57. Raynbird, 'On the Farming of Suffolk', p. 266; William Wickett and Nicholas Duval, *The Farmer's Boy: the Story of a Suffolk Poet Robert Bloomfield, 1766–1823* (Terence Dalton, Lavenham, 1971).

13 SHROPSHIRE: REALITY AND SYMBOL IN THE WORK OF MARY WEBB

John H. Paterson and Evangeline Paterson

Why should a novelist – any novelist – choose a real place, in a known area, and then use it as a setting, under the thinnest of disguises, for an imaginary cast of characters? If the imagination can stretch to the casting, why can it not cover the topography as well? Why allow oneself to be constrained from the outset by a geography dictated not by the fiction but by the reality? What advantage, if any, does this give to the novelist?

These are obvious questions, but they do not seem previously to have been asked, let alone answered. With the essentially historical novelist, of course, the constraint is unavoidable; he or she has imaginative licence only within the known framework of fact and place: the larger outcome is already on record, and the fiction must conform, at least in general outline, to the historical sequence. Even so fine and careful a novelist as Margaret Mitchell was confronted, after writing *Gone With The Wind*, with the choice that she must either adjust the date of the Battle of Gettysburg or explain, retrospectively, that a birth had been premature. But for the ordinary writer of fiction, unhampered by these larger demands of time and story, the constraint of place seems self-imposed, and unnecessarily so.

In seeking an answer to the questions posed, we can dismiss immediately some explanatory trivia. Walter Scott, himself a forerunner of the regional novelists of a later day, once commented that 'local names and peculiarities make a fictitious story look so much better in the face'.[1] That may be so, but it may equally be counter-productive, by leading the reader to expect a topographic consistency which the writer, later on, wishes to override. There is nothing so trying as a real topography redrawn, and generations of youthful enthusiasts must, like the present writer, have been baffled or enraged by Arthur Ransome's unapologetic transfer of the Old Man of Coniston to the shores of Lake Windermere in his *Swallows and Amazons* and its successors.

There seem rather to be three main answers to our question: why? It is the purpose of this chapter briefly to consider them, and to do so in relation to the work of Mary Webb, one of the most emphatically regional of all our novelists.

Landscape Reality in Fiction

The first reason why the novelist may set the story in a real landscape is because there — and perhaps only there — does he or she feel confident of presenting a familiar and ordered life. The characters in any story engage our interest and sympathy, not so much by the extraordinary as by the ordinary circumstances of their living: not by murder or mayhem but by the intrusion of these things, if intrude they must, upon a stable and credible lifestyle. On one side, then, the topographic setting acts as a *stabilising* factor if, as novelists sometimes find, their characters take on a life of their own and get out of hand. On the other, it imposes upon the writer a necessary *discipline* against his wilder flights of fancy.

In any case, the novelist writes out of his or her own experience, and that experience has, for many writers, been acquired within a geographically restricted area. This may be the *only* ordered life known to the novelist, but the corollary of that is that he or she can write of it with confidence. A part of the imaginative effort involved in the writing can be transferred away from the setting, which in this sense is a known quantity, and concentrated upon the actors.

This first reason for choosing a regional setting we might characterise as *authenticity*. It is possible, although not necessary, then, to take the argument further, and to claim that the region imparts not only realism to the story but a necessary inspiration to the writer. It was W. B. Yeats who wrote that he and his friends in the Irish renaissance thought

> All that we did, all that we said or sang
> Must come from contact with the soil, from that
> Contact everything Antaeus-like grew strong.

As it happens, the Irish writers, whether considered individually or as a group, afford this belief little support, but it was a belief which Mary Webb certainly held, as her essay 'Vis Medicatrix Naturae' shows, and therefore has a certain relevance to our present study. Away from Shropshire, it seems clear, she was less than herself.

The second reason for using a regional setting we might call *identity*. Familiar with a region and its inhabitants, the writer sets out deliberately to display it to the outside world, perhaps as an example to be shunned or followed; perhaps as a museum piece to be preserved or understood. And displayed not to the outside world only, but to its own people also, to make them more aware of what surrounds them and of the way of life they have inherited.

It is conceivable that this exercise, this establishing of regional identity, might be done as advertising, for motives of profit to the writer – conceivable, but not likely. Scott may have been the father of the Scottish tourist trade, but Lawrence can hardly have expected to draw throngs of visitors to the Nottinghamshire coalfield, and if Lewis Grassic Gibbon[2] set off a rush of tourists to the Howe of the Mearns it has gone unnoticed in eastern Scotland. For the most part, the regional novelist performs his or her task for the sake of filial duty or aesthetic pleasure alone, to create a sense of regional identity: to say, in effect, 'this region has qualities of place and people which make it special'.

The third reason for choosing real landscape as the setting for a novel may be *symbolism*. This requires some explanation. Studies in perceptual geography have, in recent years, carried us some way towards an understanding of the mental maps which we all carry in our minds, and of the principle that what we *think* is there is more important to our decision-making than what actually *is*. But there are other perceptions of great antiquity and ancient understanding, whereby the real world is symbolising something – is conveying a series of messages. Nature is a book to be read; spatial features can be, and are to be, *interpreted*. This introduces us to the mythical–magical conception of space,[3] in which the disposition and shape of things take on a significance beyond themselves, and particular places become symbols of events which have, or are believed to have, taken place there. The problem then becomes one of reading this book of nature, this *paysage moralisé*, as Mary Webb's own biographer calls it.[4]

This third use of regional landscape is not common among regional novelists, and for very obvious reasons. The danger of appearing to support an antiquated determinism is evident. Credibility on the part of the modern reader is hard to achieve, and those who do believe are unlikely to be the same people as those who read the novel. Scott appears to have viewed Scotland as a *paysage moralisé*, the Highlands symbolic of the wild, untamed (but not necessarily uncultivated – there were noble savages there) men who inhabited them; the Lowlands symbolic of art, industry and peace (see especially *Rob Roy* and the introduction to *The Fair Maid of Perth*).

Most of the better-known landscapes of symbolism are to be found in pure works of the imagination where, following the lead of medieval painters of heaven and hell (with appropriate topographic detail), they have been deliberately created to fit the story, whether it is Bunyan's *Pilgrim's Progress* or C. S. Lewis's *Narnia* tales. But Mary Webb's first novel, as we shall see, is built upon a symbolic landscape, and all her books abound in references to the superstitions of country folk about people, places and things. It is to her works that we will now turn.

Landscape in the Novels of Mary Webb

Of the three explanations for the choice of a real regional setting for an account of fictitious events — authenticity, identity and symbolism — the first can be immediately applied to our author. Born in Shropshire, Mary Webb lived successively in Leighton, Much Wenlock, Stanton-upon-Hine Heath, Meole Brace, Pontesbury and Lyth Hill, and was desperately unhappy whenever marriage or family demands removed her elsewhere. If she was ever going to write of a world she knew, then it had to be Shropshire: there was no other.

That world, however, she knew intimately, with an intimacy recognised by her friends and neighbours, and which comes through to the reader of both her prose and her poetry. Encouraged by her father, she developed as a child habits of observation and of what an earlier generation might have called communing with nature — a cliché in our own times, but one which Mary Webb herself took completely seriously, as witness her comment that 'the complete character is that which is in communion with most sides of life . . . which has for its fellows the sympathy of understanding, for nature the love that is without entire comprehension', and her quotation with approval of Sir Thomas Browne's view, 'We live the life of plants, the life of animals, the life of men, and at last the life of spirits.'[5]

Her well-attested habits included sitting motionless for hours — sometimes a whole day — in a particular spot, and scorning paths for the direct route between points, regardless of whatever vegetation might impede her. These habits bore fruit in such passages as the description of Sarn Mere which opens *Precious Bane*:

> there's a discouragement about the place. It may be the water lapping, year in and year out — everywhere you look and listen, water; or the big trees waiting and considering on your right hand and on your left; or the unbreathing quiet of the place, as if it was created but an hour gone, and not created for us . . . And there was such a dream on the place that if a wild bee came by, let alone a bumble, it startled you like a shout.[6]

Or in poems like 'Little Things':

> The tiny spiders on wych elms in May,
> Of rare pale green; the young and downy bee,
> Singing her first low song; the white ant's cradle –
> They crowd upon us with their mystery.[7]

Much of this was, of course, general to nature rather than particular to her native county, but the locational references are there, too:

> Trewern Coed [in the Clun Forest] was a typical border village, not quite sure of its nationality, mingled in speech, divided between the white, blue-roofed cottages of Wales and the red thatched ones of Shropshire. It lay in a hollow of the hills that were round it like a green-clad arm, and a broad shallow river washed its gardens.[8]

Most of Mary Webb's place-specific allusions express this same consciousness, that she lived in a border zone (shades of *The Fair Maid of Perth*) between the wild hills of Wales and the fertile lowlands of England, with 'her' country – the hills between Shrewsbury and Bishops Castle – marking a sort of divide:

> the wind raged over the vast, rolling plain in the west – a plain utterly different from that on the other side of the two ridges [the Long Mynd and the Stiperstones], which was flat except for a few round hills between the rose-coloured turnip-fields and the diaper of stubble and grass. But the smallest hills to the west would have been striking features in an ordinary countryside; the valleys were chasms, the flattest lands a switchback.[9]

To this contrast she attached, as we shall later see, a symbolism that underlay much of her thinking.

To begin with, to be sure, she was inclined to overdo the landscape description. It was the error of inexperience, and was largely confined to her first novel, *The Golden Arrow*, from whose early pages the following over-rich description is taken:

> The tesselated plain, minute in pattern as an old mosaic, seemed on this fervent day to be half-molten, ready to collapse. The stable hills shook in the heat-haze like a drop-scene just lifting upon reality. The ripening oat-fields, the already mellow wheat seemed like frail wafers prepared for some divine bacchanalia. A broad pool far down among black woods looked thick-golden, like metheglin in a small ebony cup.[10]

This is altogether too lush; every landscape detail has to remind the observer of a non-landscape simile. But it was an error which she swiftly corrected; the second of her novels, *Gone To Earth*, and the subsequent books have all the atmosphere, and the carefully reconstructed country speech, yet almost none of the descriptive passages, of the first. The players in the drama have, very properly, become more important than the stage.

We come now to the second justification for the regionally based novel, that of establishing a regional identity. That Mary Webb saw Shropshire as a region intermediate between England and Wales, and sharing some of the characteristics of each, both in landscape and people, has already been recognised. Its distinctiveness lay in the mixture. That Mary Webb's intentions were in part didactic can also be readily established from the novels, and even more clearly from her non-fiction. She wished to teach others to see with her eyes, to recognise the character of things around them, to pause and consider nature in its immense variety. That there were lessons to be learned she had no doubt:

> We need no great gifts – the most ignorant of us can draw deep breaths of inspiration from the soil . . . The primal instincts can seldom be so dead that no pleasure or kinship wakens at the thronging of these vivid colours and mysterious sounds. Here is a kingdom of wonder and of secrecy into which we can step at will.[11]

That her didactic impulse drove her further than this (and it did so) is, perhaps, unfortunate. She seems at times to have been trying to convert her readers into nature-lovers against their will. And she fell, in the beginning, into the habit of commenting editorially on the actions of her characters in a way which, by the time *The Golden Arrow* appeared in 1916, was distinctly out of fashion in the English novel, and which is liable to vex the modern reader. This is the more curious and regrettable when we consider that, in her comments on sexual matters and on relations between the sexes she was, if anything, *ahead* of her time – frank and open for the period and especially frank and open for a lady novelist of the period.

So she established a regional identity for her Shropshire homeland and its people, and she strove to teach her readers to observe and appreciate landscape. How successful was she? There can, of course, be no objective criteria by which to judge, but there can be comparisons. Compared with Hardy (whom she read and admired) she was surely no less successful in the creation of regional atmosphere: reading her novels, one is no less aware than with Hardy's of the background of land and sky — at least when, from *Gone To Earth* onwards, she had indeed placed them in the background, where they belong. Her novels have survived and attracted new readers and new editions, without (perhaps in spite of) the efforts of Stanley Baldwin to draw attention to them. We speak today of the Mary Webb country, and we know precisely where it is to be found. By these criteria, at least, she succeeded.

But we must now turn to our third topic: landscape as symbol in Mary Webb's work. On this subject her most recent biographer, Gladys Mary Coles, has written most perceptively, showing how thoroughly she understood the strong vein of superstition in the people of her country; in *Precious Bane* almost two hundred legends and superstitions can be counted.[12] In her first and last novels, *The Golden Arrow* and the unfinished *Armour Wherein He Trusted*, Mary Webb attempted the formidable task of using the topography of her settings as a symbol of the characters and their actions. As already noted, this has not often been successfully attempted in English-language fiction; Coles cites Nathaniel Hawthorne (*The Scarlet Letter* and *The Marble Faun*) as a predecessor on this treacherous path, but names no other example.

Obviously, the major obstacles to the whole operation are, first, the commitment to a particular kind of what a geographer might call environmental determinism, and the need to sustain the naturalistic parallel through thick and thin; secondly, the difficulty of finding a landscape amenable to this type of interpretation. It is a commonplace to find deities located by popular consent on the highest and most inaccessible mountain top, or spirits speaking to untutored minds from volcanoes or caves or waterfalls, but none of this is particularly appropriate to the English Midlands.

To have a *paysage moralisé* one must above all have *contrast*; otherwise the story built upon it is hardly likely to generate momentum. With Hawthorne, it was the contrast between the wilderness and the sown — the same contrast which Scott exploited across the Highland Boundary Fault in Scotland. And Mary Webb was fortunate — is there any other word one can use? — to be born into, and live her life in, an area which offered contrast at two levels, local and regional. Such skill as she showed in exploiting this fact was hers alone but, compared with other novelists, she was in this respect off to a flying start before she realised it.

The local landscape contrast which first caught her imagination, and which underlies *The Golden Arrow*, is that between the two ridges of mountain to the west of Church Stretton: the Long Mynd (which she called the Wilderhope Range) and the Stiperstones ridge (her Diafol Mountain). Topographically, the contrast between them is clear: the first, or easterly, ridge rises above the Church Stretton Fault in a series of rounded hills, concordant in summit, in places plateau-like, and covered by grass and heather moor. The westerly ridge, by contrast, is narrow and topped by a line of rocky outcrops, the most impressive of which is known as the Devil's Chair. The geological distinction between the two ridges accounts for these outlines: the Long Mynd is composed of Pre-Cambrian grits while the Stiperstones ridge (from which it is separated by a valley cut in Cambrian shales) represents the outcrop of Ordovician quartzites which form the upthrust eastern extremity of a wide area of Ordovician formations known as the Shelve district.

Around the Devil's Chair, generations of country people had woven a web of legend and the novelist, to begin with, was doing no more than record a widely felt sensation when she wrote:

> On the highest point of the bare, opposite ridge, now curtained in driving storm-cloud, towered in gigantic aloofness a mass of quartzite . . . Nothing ever altered its look. Dawn quickened over it in pearl and emerald; summer sent the armies of heather to its very foot; snow rested there as doves nest in cliffs. It remained, inviolable, taciturn, evil . . . For miles around, in the plains, the valleys, the mountain dwellings it was feared. It drew the thunder, people said . . . It was understood that only when vacant could the throne be seen. Whenever rain or driving sleet or mist made a grey shechinah there people said, 'There's harm brewing. He's in his chair.'[13]

The novelist with a Devil's Chair in his or her region is clearly possessed of a valuable property. The problem then becomes one of using that property to advantage. In her first novel, Mary Webb decided to extend or enlarge her *paysage moralisé* to embrace the whole of the two ridges, and to create parallels to the landscape in the lives of her characters. So the Long Mynd became a place of security, order and comfort, where the Flockmaster tended his stock with skill and devotion; a placid land where there was nothing to threaten — at least, not so long as the shepherd was on hand. It came to symbolise peace and virtue.

The Stiperstones ridge stood in complete contrast: 'The sheep that inhabited these hills would, so the shepherds said, cluster suddenly and stampede for no reason, if they had grazed too near it [the Devil's Chair] in the night.'[14] Its inhabitants were for the most part miners — dark, underground creatures — and its landscape symbolised the wild and wayward streak in Mary Webb's characters. Here took place events in the story which tremble on the brink of the Gothic: the daughter of the Flockmaster went to live there with the mine foreman, out of wedlock, and ended by burning down her own cottage and possessions, while the young man, tormented by passions which he could neither understand nor control, tried eventually to blow up the Devil's Chair with explosives from the mine:

> above, like a fortress on the bleak sky, loomed the Chair — unexpectant, imperturbable, sinister. Stephen loathed it. He knew all about quartzite and its enduring nature; he knew that for thousands of years the Chair had fronted everlastingness while men died like flies . . .
>   'I'll blow the old rocking chair sky high,' he said to himself, 'and see how it'll look then.'[15]

He failed, of course, and abandoned home and family for America, only to return at the end of the book for a reconciliation scene into which Mary Webb tries again to introduce a symbolic note. This time,

> It seemed to her that there was no hostility now between the two ranges, between the towering throne and the small white cross [a signpost on the Long Mynd]. Always before, she had superstitiously regarded the Chair as wholly evil, the Flockmaster's signpost as wholly good. Now she saw good and evil mingled . . .[16]

But this is the trouble with symbolic landscapes: you cannot alter the symbols half way through to suit the story — the symbolism must be consistent. There is no question of Bunyan staging a return visit to the Valley of the Shadow of Death and finding that it has been turned into a health resort. As it happens, and despite this difficulty, however, Mary Webb came remarkably close to consistency in *The Golden Arrow* — to the detriment of the story, most readers would feel. The packaging into two contrasting landscapes and groups of characters is a little too neat.

The second novel, *Gone To Earth*, continues the packaging of characters, but not of landscapes. The principal action takes place,

however, on the Stiperstones ridge, and the climax on its northern
end, Lords Hill (easily recognisable in the story as God's Little
Mountain). This forewarns us, perhaps, that the ending will be tragic,
as indeed it is, but the landscape symbolism is on this occasion not
pressed. Of the two men who pursue the heroine, Coles says, in the
introduction to the 1978 reprint of the novel, 'they are representative
of the opposing physical and spiritual values between which Hazel
swings (this gives the flavour of a Morality): the well-defined contrast
between them is fundamental to plot and theme'.[17] In other words,
the search for contrast goes on, in the sphere of character, but
without pressing further the landscape symbolism.

This being the case, it is remarkable that, in her last, unfinished
novel, Mary Webb returned to her topographic symbolism, this time
on the larger, regional scale. *Armour Wherein He Trusted* extends to
some 70 pages in the printed version of 1928, is set in the eleventh
century and represents an exceedingly bold venture for an authoress
whose most daring (and largely successful) previous exploration of the
past had been the reconstruction (in *Precious Bane*) of the life of
nineteenth-century Shropshire, depicted through the eyes of a
countrywoman of the period. But a gap of eight centuries or so is an
imaginative obstacle of an altogether different order of magnitude;
yet Mary Webb tackled it, by general critical consent, with a real
measure of success.

It is not, however, the quality of this first draft of half a novel
which concerns us here, but its setting. For the author has turned back
in her last work to those regional contrasts in which Shropshire stands
midway between two worlds, Wales and England. Here on the Marches
the two were for centuries in conflict, and the novel is full of castles
and crusading. And these two worlds symbolised for her a personal
conflict. We can follow Coles' comments here:

> From the outset in *The Golden Arrow*, Mary Webb's 'poetic or
> fairy precinct' is established, the actual geographical location being
> her hill country, the border region of south-west Shropshire, once
> part of the old Kingdom of Powys, a mid-way realm in the Marches
> of Wales where Saxon and Celt intermingle in blood and mind,
> language and myth. It is a region which she described later . . . as
> 'the country that lies between the dimpled lands of England and
> the gaunt purple steeps of Wales — half in Faery and half out of
> it . . .' This, for her, becomes a symbolical borderland between
> spiritual and material, where nature is both veil and image. Her

'land of Betwixt and Between', as she finally defined it in the last novel . . .[18]

'Half in Faery and half out of it': she seems to have been fascinated by those distant glimpses of the Welsh mountains which one obtains from 'her' hill country — by the distant presence of Cader Idris and the mystery of the unknown and (so far as we can tell) unvisited region between. Imagination must have filled the world to the west with both mystery and wonder. For her, there was a regional distinctiveness here which transcended even topography and language.

Conclusion

Mary Webb could probably be described as a mystic; perhaps as a pantheist; certainly as a woman with a passion for nature and its preservation. To describe her as a geographer in spirit would be pressing the matter altogether too far, but, in an era when humanistic geography, perception studies and topophilia all form part of a geographer's stock in trade, it is worth concluding with a comment of hers about people and places which is, in essence, not far away from what we geographers have been telling each other during the 1970s:

> For the personality of a man reacting upon the spirit of a place produces something which is neither the man nor the place, but fiercer and more beautiful than either. This third entity, born of the union, becomes a power and a haunting presence — non-human, non-material. For the mind that helped to create it once, it dominates the place of its birth for ever.[19]

Notes

1. Cited in J. H. Paterson, 'The Novelist and his Region: Scotland through the eyes of Sir Walter Scott', *Scottish Geographical Magazine*, vol. 81 (1965), p. 148.
2. G. Whittington, 'The Regionalism of Lewis Grassic Gibbon', *Scottish Geographical Magazine*, vol. 90 (1974), pp. 75–84.
3. R. D. Sack, 'Magic and Space', *Annals, Association of American Geographers*, vol. 66 (1976), pp. 309–20.
4. Gladys M. Coles, *The Flower of Light: A Biography of Mary Webb* (Duckworth, London, 1978), p. 143.
5. Mary Webb, *Spring of Joy* (Jonathan Cape, London, 1928), pp. 131–2.
6. Mary Webb, *Precious Bane* (Jonathan Cape, London, 1928), p. 1.
7. Mary Webb, 'Little Things', *Fifty-One Poems* (Jonathan Cape, London, 1946).

8. Mary Webb, *Seven For A Secret* (Jonathan Cape, London, 1928), p. 111.

9. Mary Webb, *The Golden Arrow* (Jonathan Cape, London, 1928), p. 203.

10. Ibid., p. 32.

11. Webb, *Spring of Joy*, pp. 130–1.

12. Coles, *The Flower of Light*, pp. 136 and 143–5.

13. Webb, *The Golden Arrow*, pp. 40–1.

14. Ibid., p. 41.

15. Ibid., pp. 249 and 253.

16. Ibid., pp. 340–1.

17. Gladys M. Coles, introduction to Mary Webb, *Gone To Earth* (Duckworth, London, 1978), p. 6.

18. Coles, *The Flower of Light*, p. 145.

19. Webb, *The Golden Arrow*, p. 222.

The Mary Webb novels, with publication dates, are: *The Golden Arrow* (1916); *Gone To Earth* (1917); *The House in Dormer Forest* (1920); *Seven For A Secret* (1922); *Precious Bane* (1924); *Armour Wherein He Trusted*, published in *Spring of Joy* (1928).

NOTES ON CONTRIBUTORS

Howard F. Andrews	Associate Professor of Geography, University of Toronto.
Ian G. Cook	Lecturer in Geography, Liverpool Polytechnic.
Denis Cosgrove	Principal Lecturer in Geography, Oxford Polytechnic.
William J. Lloyd	Assistant Professor of Geography, University of Texas at El Paso.
Catherine A. Middleton	Research Student in Geography, University College, London.
Peter T. Newby	Lecturer in Geography, Middlesex Polytechnic.
Gunnar Olsson	Professor of Geography and Planning, Nordic Institute of Planning, Stockholm.
Kenneth R. Olwig	Post-doctoral Fellow in Geography, Royal Danish Graduate School of Education Studies, Copenhagen.
Evangeline Paterson	Poet and editor.
John H. Paterson	Professor of Geography, University of Leicester.
Douglas C. D. Pocock	Senior Lecturer in Geography, University of Durham.
Hugh C. Prince	Reader in Geography, University College, London.
Christopher L. Salter	Associate Professor of Geography, University of California at Los Angeles.
David Seamon	Research Fellow, Department of Social and Economic Geography, University of Lund.
John E. Thornes	Lecturer in Geography, University College, London.

INDEX